THE FINITE ELEMENT METHOD IN CHARGED PARTICLE OPTICS

THE KLUWER INTERNATIONAL SERIES IN ENGINEERING AND COMPUTER SCIENCE

THE FINITE
ELEMENT METHOD
IN CHARGED
PARTICLE OPTICS

by

Anjam Khursheed
National University of Signapore, Singapore

KLUWER ACADEMIC PUBLISHERS
Boston / Dordrecht / London

Distributors for North, Central and South America:
Kluwer Academic Publishers
101 Philip Drive
Assinippi Park
Norwell, Massachusetts 02061 USA
Telephone (781) 871-6600
Fax (781) 871-6528
E-Mail <kluwer@wkap.com>

Distributors for all other countries:
Kluwer Academic Publishers Group
Distribution Centre
Post Office Box 322
3300 AH Dordrecht, THE NETHERLANDS
Telephone 31 78 6392 392
Fax 31 78 6546 474
E-Mail <orderdept@wkap.nl>

 Electronic Services <http://www.wkap.nl>

Library of Congress Cataloging-in-Publication Data

Khursheed, Anjam.
   The finite element method in charged particle optics / by Anjam
Khursheed.
     p.   cm. -- (The Kluwer international series in engineering and
computer science ; 519)
   Includes bibliographical references and index.
   ISBN 0-7923-8611-6 (acid-free paper)
   1. Electron optics. 2. Finite element method. 3. Particle beams.
4. Electron beams. I. Title. II. Series: Kluwer international
series in engineering and computer science ; SECS 519.
QC793.5.E62K48   1999
537.5'6--dc21

                                 99-40720
                                     CIP

Printed on acid-free paper.

Printed in the United States of America

CONTENTS

PREFACE   ix

ACKNOWLEDGEMENTS   xiii

1. FIELD THEORY   **1**

    1. **Electrostatics**   1
        1.1 The Conservative Property   1
        1.2 Gauss's Law and Poisson's Equation   2
        1.3 Orthogonal Coordinate Systems   5
        1.4 Boundary Conditions   8
        1.5 Electrostatic Energy   9

    2. **Magnetostatics**   11
        2.1 Basic Laws   11
        2.2 Boundary Conditions   12
        2.3 The Vector Potential   12
        2.4 The Vector Potential in two dimensions 13
        2.5 The Magnetic Scalar Potential   17
        2.6 The Reduced Scalar Potential   18
        2.7.The Two Scalar Potential approach   20
        2.8 The Vector Potential in three dimensions 22
        2.9 Magnetic Energy   23

2. FIELD SOLUTIONS FOR CHARGED PARTICLE OPTICS **25**

    1. **The Equations of motion**   25

    2. **The Paraxial Equation of Motion**   29

    3. **On-axis Lens Aberrations**   32

    4. **Electrostatic and Magnetic Deflection Fields** 37
        4.1 The Deflector Layout   37
        4.2 Harmonic expansions of the scalar potential 39
        4.3 Source term for Toroidal Coils   40
        4.4 Source term for Saddle Coils   42
        4.5 Boundary conditions on Multipole Electrodes 43

3. **THE FINITE DIFFERENCE METHOD**   **45**

    1. **Local finite 5pt difference equations**   46

    2. **The Matrix Equation**   47

    3. **Truncation errors**   49

    4. **Asymmetric stars**   50

    5. **Material Interfaces**   53

    6. **The nine pointed star in rectilinear coordinates**   55

    7. **Axisymmetric cylindrical coordinates**   56

4. **FINITE ELEMENT CONCEPTS**   **61**

    1. **Finite Elements in one dimension**   62
        1.1 The weighted residual approach   62
        1.2 The variational approach   66

    2. **The Variational method in two dimensions**   70
        2.1 Square elements   70
        2.2 Rectangular elements   72
        2.3 Right-angle triangle elements   73

    3. **First-order shape functions**   75
        3.1 Shape functions in one dimension   76
        3.2 Shape functions in two dimensions   77
        3.3 Example of right-angle triangle   79
        3.4 Quadrilateral element shape functions   82

    4. **The Galerkin Method**   84

    5. **Nodal equations and Matrix Assembly**   87

    6. **Axisymmetric Cylindrical Coordinates**   92
        6.1 The general nodal equation   92
        6.2 First-order elements   92
        6.3 On-axis nodes   94
        6.4 Near-axis r^2 correction   95

    7. **Edge elements**   96

5. **HIGH-ORDER ELEMENTS**   **99**

    1. **Triangle elements**   99

    2. **Quadrilateral elements**   105

    3. **The Serendipity family of elements**   106

6. ELEMENTS IN THREE DIMENSIONS 111

  1. Element shape functions 112
    1.1 The Brick element 112
    1.2 The Tetrahedral element 113
    1.3 Prism elements 113

  2. Generating tetrahedral elements to fit curved boundary surfaces 114

7. FEM formulation in Magnetostatics 125

  1. Magnetic vector potential 125
    1.1 Two dimensional field distributions 125
    1.2 Three dimensional field distributions 127

  2. The magnetic scalar potential in three dimensions 131
    2.1 The weighted residual approach 131
    2.2 The Reduced scalar potential formulation 132
    2.3 The two scalar potential formulation 133

  3. Saturation Effects 135

8. ELECTRIC LENSES 141

  1. Accuracy issues 141
    1.1 The two-tube lens example 141
    1.2 First-order and second-order elements 142
    1.3 Cubic elements 152

  2. Direct ray tracing using off-axis mesh node potentials 154
    2.1 Direct tracing vs perturbation methods 154
    2.2 Trajectory integration errors 156
    2.3 Field interpolation errors 157
    2.4 Truncation errors 159

9. MAGNETIC LENSES 161

  1. Accuracy issues 162
    1.1 The First-order approximation 162
    1.2 The Solenoid example 164
    1.3 Axial field errors for first and second-order elements 165
    1.4 Cubic elements 173

  2. Magnetic axial field continuity tests 176
    2.1 First-order elements on a trial region block mesh 176
    2.2 Mesh refinement 183
    2.3 Direct ray tracing from axial field distributions 185

  3. Magnetic field computations in three dimensions 187

10. DEFLECTION FIELDS   191

   1. Finite element formulation   191
       1.1 Energy functional   191
       1.2 Element nodal equations   193

   2. Accuracy tests   195
       2.1 The analytical solution for magnetic deflectors in free space   195
       2.2 The analytical solution for a magnetic core deflection test
            Example   197
       2.3 FEM results for magnetic deflector test examples   199
       2.4 Electrostatic deflection   204

11. MESH RELATED ISSUES   209

   1. Structured vs unstructured   209

   2. The Boundary-fitted coordinate method   217

   3. Mesh refinement for electron gun simulation   224

   4. High-order interpolation   230
       4.1 C^1 triangle interpolant   231
       4.2 Normalised Hermite-Cubic interpolation   232
       4.3 Laplace polynomial interpolation   234

   **5. Flux line refinement for three dimensional electrostatic
       problems**   238

   6. Accuracy tests   239
       6.1 High-order interpolation in two dimensions   239
       6.2 The Ampere Circuital mesh test   248
       6.3 High-order interpolation in three dimensions   253

Appendix 1: Element Integration formulas   257
       1. Gaussian Quadrature   257
       2. Triangle elements   258

Appendix 2: Second-order 9 node rectangle element pictorial stars   259

Appendix 3: Green's Integration formulas   265

Appendix 4: Near-axis analytical solution for the solenoid test example   267

Appendix 5: Deflection fields for a conical saddle yoke in free space   269

INDEX   273

PREFACE

In the span of only a few decades, the finite element method has become an important numerical technique for solving problems in the subject of charged particle optics. The situation has now developed up to the point where finite element simulation software is sold commercially and routinely used in industry.

The introduction of the finite element method in charged particle optics came by way of a PHD thesis written by Eric Munro at the University of Cambridge, England, in 1971 [1], shortly after the first papers appeared on its use to solve Electrical Engineering problems in the late sixties. Although many papers on the use of the finite element method in charged particle optics have been published since Munro's pioneering work, its development in this area has not as yet appeared in any textbook. This fact must be understood within a broader context. The first textbook on the finite element method in Electrical Engineering was published in 1983 [2]. At present, there are only a handful of other books that describe it in relation to Electrical Engineering topics [3], let alone charged particle optics. This is but a tiny fraction of the books dedicated to the finite element method in other subjects such as Civil Engineering. The motivation to write this book comes from the need to redress this imbalance.

There is also another important reason for writing this book. The development of the finite element method in charged particle optics is relatively unknown amongst the wider community of finite element researchers. This is unfortunate, since the finite element method in charged particle optics has inspired unique developments of its own, and some of these may well be beneficial to other subjects.

This book aims at providing an introduction of the finite element method in charged particle optics to the general reader. It goes beyond merely summarising previous work, and attempts to place the finite element method within a wider framework. Examples of mesh generators and extrapolation methods for the finite element method developed in fluid flow and structural analysis, not hitherto used in charged particle optics will be presented. Detailed test examples are given and it is hoped that they may serve as bench mark tests and be a way of comparing the finite element method to other field solving methods. The book reports on some high-order interpolation techniques and mesh generation methods developed within charged particle optics which may be of interest to other finite element researchers.

No attempt here is made to describe the growing number of programs being developed for charged particle optics applications. This is being done elsewhere [4].

The subject of charged particle optics covers a wide area of research that ranges from the design of particle accelerators, scanning electron microscopes through to the simulation of television tubes. It is important to state from the outset that the kind of finite element developments reported here only relate to the design of instruments like electron microscopes. This naturally means that subjects such as calculating high frequency (RF) field distributions are not described. This book is primarily concerned with the deflection, guidance and focussing of charged particle beams by static electric or magnetic fields.

The material covered in this book is intended to serve as graduate text, suitable for postgraduate students and researchers working in the subject of charged particle optics and other finite element related applications.

An introduction to the kinds of field problems that occur in charged particle beam systems is presented in Chapters 1 and 2. These chapters are primarily intended for the non-specialist in charged particle optics.

Chapter 3 summarises some aspects of the finite difference method, which are closely related to the finite element method. Chapter 4 outlines finite element theory and procedure. Starting with simple examples in one dimension, it shows how partial differential equations can be cast into integral form and made suitable for finite element representation. An introduction to the local polynomial function expansions used for each element, known as shape functions is given. The detailed formulation of local and global matrices is also presented with simple examples.

Chapter 5 covers the important subject of higher-order elements, which can be an effective way of improving finite element accuracy, while Chapter 6 describes the finite element method in three dimensions. Different ways to formulate scalar and vector problems for magnetic fields are presented in Chapter 7.

Some test field distributions are examined to assess the accuracy of finite element solutions in Chapters 8 and 9. They show how high-order elements and extrapolation methods can significantly reduce truncation errors.

Chapter 10 deals with deflection fields, while Chapter 11 describes various developments in mesh related issues. This latter chapter demonstrates how mesh generation and refinement is an integral part to finite element accuracy.

ANJAM KHURSHEED

Singapore

REFERENCES

[1]   E. Munro, Computer-Aided Design Methods in Electron Optics, PHD Thesis, University of Cambridge, England, October 1971

[2]   P. P. Silvester and R. L. Ferrari, "Finite Elements for Electrical Engineers", Cambridge University Press, Cambridge, England, 1983.

[3]   See for instance, S. Ratnajeevan H. Hoole, "Computer-aided analysis and Design of Electromagnetic Devices", Elsevier, London, 1989, and Jianming Jin, "The Finite Element Method in Electromagnetics", John Wiley and Sons, Inc., New York, 1993 and K. J. Binns, P. J. Lawrenson and C. W. Trowbridge, "The Analytical and Numerical Solution of Electric and Magnetic Fields", John Wiley and Sons, New York, 1992

[4]   A research group under Professor P. Kruit in the Applied Physics Department of Delft University of Technology, Delft Holland, maintain a charged particle optics software internet service. They have hosted various conferences on the current software available. Further information can be obtained from Professor P. Kruit, Vakgroep DO, Lorentzweg 1, 2628 CJ Delft, The Netherlands.

ACKNOWLEDGEMENTS

A book of this nature usually relies on contributions from other people, and this book is no exception. I am grateful to: Zhao Yan, my PHD student, for work relating to deflection fields; to Zhu Aiming, my Masters student, for simulations using direct ray tracing; to my project students, Lee Meng Khuan for work on high-order elements, and Wong Heng Loon for developing some aspects of three-dimensional mesh generation.

On a broader note, I would like to dedicate this book to four people, who in their different ways, have shaped my life and career. The first is Alan Russell Dinnis, who supervised my PHD from 1979 to 1983 in the Electrical Engineering Department at the University of Edinburgh, Scotland. Alan quickly became my colleague and friend, and through a period of over fifteen years, we published many papers together and collaborated on international projects. He had the rare quality of trusting his PHD students, and treating them as his equals. He would politely correct their mistakes and at the same time be encouraging. He was one of the few engineers I knew who could understand electromagnetics and electron optics. Many of the ideas presented in this book were conceived in conversation with him. He provided the experimental test ground for my theoretical work, and for this, I am always in his debt. On the human side, we enjoyed ourselves. I not only had the support of Alan, but also of his wife Kirsti and three daughters. We were all often together at international conferences, and their friendship will always be with me. While this book was being written, Alan died of cancer. This book is a living tribute to his memory.

The second person to whom this book is dedicated is my mother, Rukhsana. She also died of cancer while this book was being written. If I write a hundred books and dedicate each one to her, that will not even approach a tiny fraction of the gratitude I feel towards her.

The last two people to whom I would like to dedicate this book are my wife, Antonella, and my daughter Sarah. Antonella's contribution goes far beyond general encouragement. She type-set the whole manuscript, arranged it, and shaped it into camera-ready format. I am grateful for her help and have greatly enjoyed working with her. To Sarah, my five-year old daughter, the dedication merely expresses my good wish for her future.

ANJAM KHURSHEED

Chapter 1
FIELD THEORY

A short introduction to field theory is given here primarily for the general reader not familiar with charged particle optics or electromagnetic theory. Since the finite element method solves partial differential equations, it is first necessary to understand the conservation laws that they are founded upon and study the different ways that these equations can be formulated.

1.   ELECTROSTATICS

1.1   The Conservative Property

From the definition of the electric field $\mathbf{E}$, the force $\mathbf{F}$ acting on a charge q is

$$\mathbf{F} = q\mathbf{E} \tag{1.1}$$

The work done ΔW in moving this charge along a distance $\mathbf{dl}$, is therefore

$$\Delta W = -q\int \mathbf{E} \cdot \mathbf{dl} \tag{1.2}$$

The work done is closely related to electric potential. The change in electric potential Δu is defined as

$$\Delta u = -\int \mathbf{E} \cdot \mathbf{dl} \tag{1.3}$$

so the work done on the charge is given by

$$\Delta W = q\Delta u \tag{1.4}$$

The conservative property of the electrostatic field means that it is the difference in potential at the end points of the path that determines the work done. If a particle moves from point A at potential u_A to point B which is at potential u_B, the work done on the particle is

$$\Delta W = -q \int_{A}^{B} \mathbf{E} \cdot \mathbf{dl} = q(u_B - u_A)$$ (1.5)

The precise path that a charged particle takes through the electrostatic field does not affect the change in energy experienced by the particle. This conservative property of charged particle motion through an electrostatic field is of great importance in charged particle optics. If an electron is emitted from a 0 volt surface and accelerated to 1000 volts, it gains 1keV in energy, irrespective of its trajectory path. In situations where the path of the charged particle returns to its initial position, then obviously,

$$\oint \mathbf{E} \cdot \mathbf{dl} = 0$$ (1.6)

The conservative property of the electrostatic field states that there is no net work done on a charged particle if it returns to its initial position.

1.2   Gauss's Law and Poisson's Equation

The differential equation that describes electric field distributions is in effect, the point to point enforcement of Gauss's Law, which may be stated in the following way

$$\iint_{S} \mathbf{D} \cdot \mathbf{dS} = Q$$ (1.7)

where $\mathbf{D}$ is the electric flux density, $\mathbf{dS}$ represents the vector normal to a surface S, as shown in Figure 1.1 and Q is the total charge enclosed by the surface S,

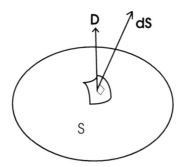

Figure 1.1 Electric flux $\mathbf{D}$ crossing a surface S

Gauss's law is the mathematical statement of the conservation of electric flux principle. If Q is zero, it states that the net flux crossing a closed domain, given by the surface S in this case, must be zero. For the more general case, it states that the net flux crossing the surface S is directly related to the source charge contained within the volume bounded by S. The source charge supplies and sustains the electric flux.

Consider the small rectangle element in two dimensions shown in Figure 1.2

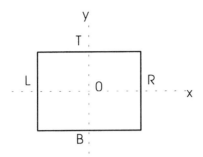

Figure 1.2 Small element in Cartesian coordinates

Let the size of this rectangle be Δx by Δy, and let the source contained within it be dQ. Gauss' law taken around this element is given by

$$-\int_L D_L dl + \int_R D_R dl + \int_T D_T dl - \int_B D_B dl = dQ \qquad (1.8)$$

where D_L, D_R, D_B and D_T are the flux densities on the sides L, R, B and T respectively. Here, the directions normal to each side are taken into account. At this point, an approximation needs to be made about how the electric flux density varies within the element. For a first order approximation with respect to the electric flux density components at the cell centre, D_{0x} and D_{0y}, are obviously given by

$$D_L \approx D_{0x} - \left(\frac{\Delta x}{2}\right)\left(\frac{\partial D_{0x}}{\partial x}\right) \quad D_R = D_{0x} + \left(\frac{\Delta x}{2}\right)\left(\frac{\partial D_{0x}}{\partial x}\right)$$

$$D_T \approx D_{0y} + \left(\frac{\Delta y}{2}\right)\left(\frac{\partial D_{0y}}{\partial y}\right) \quad D_B = D_{0y} - \left(\frac{\Delta y}{2}\right)\left(\frac{\partial D_{0y}}{\partial y}\right) \qquad (1.9)$$

The total flux in the x direction is then given by

$$-\int_L D_L dy + \int_R D_R dy \approx \Delta x \Delta y \frac{\partial D_{0x}}{\partial x} \qquad (1.10)$$

Likewise, the total flux in the y direction can be approximated to be

$$-\int_B D_B dx + \int_T D_T dx \approx \Delta x \Delta y \frac{\partial D_{0y}}{\partial y} \tag{1.11}$$

Gauss's law for the rectangle is then given by

$$\frac{\partial D_{0x}}{\partial x} + \frac{\partial D_{0y}}{\partial y} \approx \frac{dQ}{\Delta x \Delta y} \approx \rho(x, y) \tag{1.12}$$

where $\rho(x,y)$ is the charge density at point (x,y).

So long as the rectangle has finite dimensions, the above equation is only an approximation to Gauss's law. But if now, we let Δx, $\Delta y \rightarrow 0$, then the above equation becomes exact, that is, for the point (x,y), the divergence of the electric flux density should always be strictly equal to the source charge density at that point

$$\nabla \cdot \mathbf{D} = \rho \quad \forall x, y \tag{1.13}$$

This is Poisson's equation, true for all orthogonal coordinate systems. Note that it takes into account point to point variations of the dielectric permittivity $\varepsilon(x,y)$. In terms of the electric potential $u(x,y)$, where $\mathbf{D}=-\varepsilon \nabla u$, then

$$\nabla \cdot (\varepsilon \nabla u) = -\rho \tag{1.14}$$

For situations where the dielectric permittivity is constant

$$\nabla^2 u = \frac{\partial^2 u}{\partial x^2} + \frac{\partial^2 u}{\partial y^2} = f(x, y) \tag{1.15}$$

and for no source charge, this equation becomes Laplace's equation

$$\nabla^2 u = \frac{\partial^2 u}{\partial x^2} + \frac{\partial^2 u}{\partial y^2} = 0 \tag{1.16}$$

The partial differential equation, that is Poisson's equation, together with the boundary conditions, constitutes the complete mathematical statement of Gauss's law in infinitely small differential terms. There are two types of boundary conditions considered in this book, as shown in Figure 1.3. Firstly there is the explicit boundary condition, also known as the Dirichlet boundary condition, where the potential on the boundary Γ_A is fixed, and secondly there is a constraint on the normal derivative of the potential on boundary Γ_B. Stated in mathematical form

$u(x,y)=u_0$ on Γ_A Explicit Boundary Condition

$\dfrac{\partial u}{\partial n} = au + b$ on Γ_B Derivative Boundary Condition   (1.17)

where a and b are constants.

For the special case where $\partial u/\partial n=0$, the derivative boundary condition is usually known as the Neumann Boundary condition.

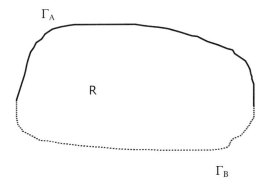

Figure 1.3 Boundary Conditions

Analytical field solutions find u(x,y) at every point (x,y) in the domain R. Numerical methods estimate the value of the potential at a finite number of selected points in the domain, usually referred to as "nodes".

Gauss's law on a mesh cell, such as the one shown in Figure 1.2, corresponds to something physically meaningful, and is in fact, a good starting pointing for numerical methods. But expressions based upon it are always approximate. On the other hand, Poisson's equation, although exact, is physically meaningless, since it involves letting Δx, $\Delta y \rightarrow 0$. Poisson's equation is strictly valid only at points. Where analytical solutions exist, they will be valid for an infinite number of such points. Poisson's equation is hypothetical, rather like the concept of a perfect circle: although it does not describe something physical, it is nevertheless, very useful. This is in a nutshell the difference between partial differential equations and the numerical methods used to solve them. The irony is however, that the derivation of a partial differential equation usually starts off by representing a conservation law in physical (finite) terms, and then goes on to make it unphysical, while numerical methods always stay in the realm of the "finite".

1.3   Orthogonal Coordinate Systems

To understand how Poisson's equation changes according to different coordinate systems, it is necessary to return to the realm of "finite mathematics". Take for instance polar coordinates (r,θ), where r is the radius and θ is the angle relative to the x axis. Here, a small element takes on the sector shape depicted in Figure 1.4, whose sides have lengths Δr, $r\Delta\theta$ and $(r+\Delta r)\Delta\theta$.

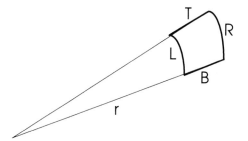

Figure 1.4 Small element in polar coordinates

The length along the side L is $r\Delta\theta$, $(r+\Delta r)\Delta\theta$ on side R, and Δr along the T and B sides. Clearly, Gauss's law for the conservation of flux applies to this element, as it does for the previous rectangular element. In this case however, the length of the sides L and R change with radius, so,

$$-\int_L D_L dl + \int_R D_R dl = -r\Delta\theta D_L + (r + \Delta r)\Delta\theta D_R \qquad (1.18)$$

The flux in the angular direction is independent of the angle, so,

$$-\int_B D_B dl + \int_T D_T dl = -\Delta r D_B + \Delta r D_T \qquad (1.19)$$

Making first order approximations for D_L, D_R, D_B and D_T in terms of the components D_{0r} and $D_{0\theta}$ at the centre of the element, and neglecting higher order terms, then,

$$\int_L D_L dl + \int_R D_R dl \approx \Delta r\Delta\theta D_{0r} + r\Delta\theta\Delta r \frac{\partial D_{0r}}{\partial r} \qquad (1.20)$$

and

$$\int_B D_B dl + \int_T D_T dl \approx \Delta r\Delta\theta \frac{\partial D_{0\theta}}{\partial\theta} \qquad (1.21)$$

Substituting these expressions into the conservation of flux equation, and dividing by the area of the element $r\Delta r\Delta\theta$, gives the following equation

$$\frac{D_{0r}}{r} + \frac{\partial D_{0r}}{\partial r} + \frac{1}{r}\frac{\partial D_{0\theta}}{\partial\theta} \approx \frac{dQ}{r\Delta r\Delta\theta} \approx \rho(r,z) \qquad (1.22)$$

As Δr, $\Delta\theta \rightarrow 0$, this equation becomes an exact point to point enforcement of Gauss's law in the domain for any (r,θ).

Poisson's equation involving spatial derivatives of the electric potential u(r,θ) are found by considering the following discrete expressions

$$D_{0r} = -\varepsilon\left[\frac{\Delta u}{\Delta r}\right] \rightarrow -\varepsilon\frac{\partial u}{\partial r}$$

$$D_{0\theta} = -\varepsilon\left[\frac{\Delta u}{r\Delta\theta}\right] \rightarrow -\varepsilon\frac{1}{r}\frac{\partial u}{\partial\theta} \tag{1.23}$$

$$\Delta r, \Delta\theta \rightarrow 0$$

When substituted into the divergence equation (1.22), this gives

$$\frac{\varepsilon}{r}\frac{\partial u}{\partial r} + \frac{\partial}{\partial r}\left(\varepsilon\frac{\partial u}{\partial r}\right) + \frac{1}{r^2}\frac{\partial}{\partial\theta}\left(\varepsilon\frac{\partial u}{\partial\theta}\right) = -\rho \tag{1.24}$$

In cylindrical coordinates, (r, φ, z), where the azimuthal angle φ is taken relative to the x axis, it is straightforward to show that the following equation enforces the conservation of flux point to point

$$\frac{D_{0r}}{r} + \frac{\partial D_{0r}}{\partial r} + \frac{1}{r}\frac{\partial D_{0\theta}}{\partial\phi} + \frac{\partial D_{0z}}{\partial z} = \rho \tag{1.25}$$

or

$$\frac{1}{r}\frac{\partial}{\partial r}\left(r\frac{\partial u}{\partial r}\right) + \frac{1}{r^2}\frac{\partial}{\partial\phi}\left(\frac{\partial u}{\partial\phi}\right) + \frac{\partial}{\partial z}\left(\frac{\partial u}{\partial z}\right) = -\frac{\rho}{\varepsilon} \tag{1.26}$$

For simplicity, the dielectric permittivity is assumed to be constant. Similarly, in spherical coordinates (r, θ, φ), where θ is the angle that a point makes with the x-y plane, and φ is the azimuthal angle in the x-y plane, it can be shown that

$$\frac{2D_{0r}}{r} + \frac{\partial D_{0r}}{\partial r} + \frac{1}{r\sin\theta}\frac{\partial(\sin\theta D_{0\theta})}{\partial\theta} + \frac{1}{r\sin\theta}\frac{\partial D_{0\phi}}{\partial\phi} = \rho \tag{1.27}$$

or

$$\frac{1}{r^2}\frac{\partial}{\partial r}\left(r^2\frac{\partial u}{\partial r}\right) + \frac{1}{r^2\sin\theta}\frac{\partial}{\partial\theta}\left(\sin\theta\frac{\partial u}{\partial\theta}\right) + \frac{1}{r^2\sin^2\theta}\frac{\partial^2 u}{\partial\phi^2} = -\frac{\rho}{\varepsilon} \tag{1.28}$$

An important theorem that comes from the conservation of flux point to point is the Divergence theorem. Here the charge Q enclosed by a surface S is replaced by a volume integral over the charge density ρ(x,y,z), so

$$\iint_S \mathbf{D}\cdot\mathbf{dS} = Q = \iiint_R \rho\,dxdydz \tag{1.29}$$

substituting now for the charge density in terms of the divergence of the electric flux from Poisson's equation gives

$$\iint_S \mathbf{D} \cdot \mathbf{dS} = \iiint_R (\nabla \cdot \mathbf{D}) dx dy dz \qquad (1.30)$$

This equation relates the net flux crossing the boundary S to the point to point conservation of flux inside the domain.

1.4   Boundary Conditions

Consider an interface between two different dielectric material types, as shown in Figure 1.5. Let the dielectric permittivity of region 1 be ε_1, and ε_2 denote the dielectric permittivity of region 2.

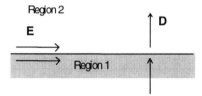

Figure 1.5 Two regions of different permittivities

From the conservative property of the electrostatic field, it is easy to show that the tangential component of the electric field must be continuous. From the application of Gauss's law it follows that the normal component of the electric flux density must be continuous. The boundary conditions can then be stated as follows

$$E_{t1} = E_{t2} = \frac{D_{t1}}{\varepsilon_1} = \frac{D_{t2}}{\varepsilon_2} \qquad D_{n1} = D_{n2} = \varepsilon_1 E_{n1} = \varepsilon_2 E_{n2} \qquad (1.31)$$

where E_{t1} , E_{n1} , D_{t1} , D_{n1} are the tangential and normal components of the electric field and flux density in region 1, and E_{t2} , E_{n2} , D_{t2} , D_{n2} are the tangential and normal components of the electric field and flux density in region 2. Note that since the tangential electric field component is continuous, the tangential component of the flux density is discontinuous. Likewise, since the normal component of the flux density is continuous, the normal component of the electric field is discontinuous.

For the special case of a perfect metal at region 1, the tangential electric field will be zero, that is, at the surface of the metal, the electric field vector will always be normal to the conductor surface. Moreover from Gauss's law, charge on the conductor surface is required to sustain the presence of this electric field. This charge density ρ_s is given by

$$\rho_s = \varepsilon_2 E_{n2} \tag{1.32}$$

These boundary conditions imply that equipotential lines will always be tangential to lossless conductors, while flux lines should always be normal to them. Also, flux and equipotential lines are expected to change discontinuously at the boundary between two different dielectric materials. Checking whether the boundary conditions are satisfied in this way often provides a quick visual check on numerically solved field distributions.

1.5   Electrostatic Energy

The stored energy in an electrostatic field is usually presented in several different ways. Since the force on a charge q in an electric field $\mathbf{E}$ is equal to $q\mathbf{E}$, the energy dW_E required to move it over a distance dl is

$$dW_E = -q\mathbf{E} \cdot \mathbf{dl} \tag{1.33}$$

If the path is taken from infinity and ends within the domain R at the point (x,y,z), then the work required to move the charge to this point can be related to the potential u(x,y,z) at that point

$$W_E = -q \int \mathbf{E} \cdot \mathbf{dl} = qu(x, y, z) \tag{1.34}$$

It can be readily shown that for n discrete charges, having charge q_i and potential u_i, the mutual interaction of the charges produces an energy

$$W_E = \frac{1}{2} \sum_{i=1}^{n} q_i u_i \tag{1.35}$$

The factor of one half arises from the fact that the mutual interaction between any pair of charges is counted twice in the summation. Physically, this corresponds to assuming, all the other charges are in position while a charge is brought from infinity to point (x,y,z). For a continuous charge distribution $\rho(x,y,z)$ over the domain R bounded by the surface S, W_E is given by

$$W_E = \frac{1}{2} \iiint\limits_R u\rho \, dR \tag{1.36}$$

Often it is more convenient to express the stored electrostatic energy purely in terms of the potential distribution. The charge distribution in the integral can of course be replaced by an expression involving the electric potential by using Poisson's equation, so

$$W_E = \frac{1}{2} \iiint_R u\rho dR = -\frac{1}{2} \iiint_R u \nabla \cdot (\varepsilon \nabla u) dR \qquad (1.37)$$

This expression can be further simplified. To do this, first consider the general vector property

$$\nabla \cdot (f\mathbf{A}) = f(\nabla \cdot \mathbf{A}) + \mathbf{A} \cdot \nabla f \qquad (1.38)$$

where f is a scalar function, f(x,y,z), and $\mathbf{A}$ is a vector. This property can be rearranged to give

$$f(\nabla \cdot \mathbf{A}) = \nabla \cdot (f\mathbf{A}) - \mathbf{A} \cdot \nabla f \qquad (1.39)$$

Now let f=u and $\mathbf{A}=\varepsilon\nabla u$, so

$$u\nabla \cdot (\varepsilon \nabla u) = \nabla \cdot (u\varepsilon \nabla u) - \varepsilon \nabla u \cdot \nabla u \qquad (1.40)$$

Forming a volume integral over the region R, this equation becomes

$$\iiint_R u\nabla \cdot (\varepsilon \nabla u) dR = \iiint_R \nabla \cdot (u\varepsilon \nabla u) dR - \iiint_R \varepsilon \nabla u \cdot \nabla u dR \qquad (1.41)$$

We can then use the divergence theorem, where

$$\iiint_R \nabla \cdot (u\varepsilon \nabla u) dR = \iint_S u\varepsilon \nabla u \cdot \mathbf{dS} = \iint_S u\varepsilon \frac{\partial u}{\partial n} dS \qquad (1.42)$$

where $\partial u/\partial n$ is the normal derivative of the potential at the boundary S. This follows from the fact that the vector $\mathbf{dS}$ is defined to be perpendicular to the boundary S.

Substituting the above relations into the expression for the electrostatic energy gives

$$W_E = \frac{1}{2} \iiint_R \varepsilon |\nabla u|^2 dR - \iint_S u\varepsilon \frac{\partial u}{\partial n} dS \qquad (1.43)$$

Many introductory textbooks to electrostatics omit the surface term by extending the boundary S to infinity where the potential is assumed to be zero. But in the context of numerical methods, this is too restrictive. The stored energy within the domain R bounded by the surface S has contributions that come from both within the domain and from the boundary itself. There are two common instances where the surface term disappears. The first situation is where u is given explicitly on the boundary, in which case the surface boundary contribution is omitted and its effect is taken into account by fixed terms in the volume integral. The second situation is where the normal derivative is zero everywhere on the boundary, that is, a Neumann boundary condition exists.

The above integral formulation is particularly useful when considering the Finite Element method. When the foregoing derivation is generalised to apply to two different functions, it is then known as Green's first theorem (given in Appendix 3). Let $v(x,y,z)$ and $u(x,y,z)$ be two scalar functions. Following a similar line of reasoning to the one just used, it is straightforward to show that

$$\iiint_R v\nabla \cdot (\varepsilon\nabla u)dR = -\iiint_R \varepsilon\nabla v \cdot \nabla u dR + \iint_S v\varepsilon\frac{\partial u}{\partial n}dS \qquad (1.44)$$

This integral equation is used in the weighted residual formulation of the finite element method.

2   MAGNETOSTATICS

2.1   Basic Laws

There are two integral equations which govern magnetostatic fields. The first is *Ampere's circuital law*, which states that the line integral of the Magnetic field intensity vector, **H**, taken around a closed loop, is equal to the enclosed current **I**

$$\oint \mathbf{H}\cdot\mathbf{dl} = I \qquad (1.45)$$

Positive current is defined as flowing in the direction of advance of a right-handed screw turned in the direction in which the closed path is traversed. Ampere's law is always true, irrespective of whether there are magnetic materials in the domain or not. A point to point enforcement of Ampere's circuital law leads to the following differential equation

$$\nabla \times \mathbf{H} = \mathbf{J} \qquad (1.46)$$

where **J** is the current density vector. This equation makes the magnetic field a non-conservative one. In electrostatics, the differential equation corresponding to the line integral of the electric field is always equal to zero. The magnetic flux density, **B**, is related to **H** by $\mathbf{B}=\mu\mathbf{H}$, where μ is the magnetic permeability.

The second integral equation of magnetostatics applies to the flux density vector **B**. It states that the magnetic flux crossing an enclosed boundary must be equal to zero, that is

$$\iint_S \mathbf{B}\cdot\mathbf{dS} = 0 \qquad (1.47)$$

This equation is based upon the conservation of magnetic flux, and is similar to Gauss's law for electrostatics. It is straightforward to show that the point to point application of the conservation of magnetic flux, like the situation for electric flux, leads to the divergence of the flux density being zero, that is

$$\nabla \cdot \mathbf{B} = 0 \qquad (1.48)$$

Unlike the situation in electrostatics however, the flux lines in a magnetic field are always continuous: that is they do not terminate on sources. The divergence of B is zero everywhere, even at the current source.

In most magnetostatic applications, saturation effects are important. This situation arises when the permeability is dependent on the field strength $\mathbf{B}$, $\mu(\mathbf{B})$, as it is for many magnetic materials. The effect of this is to make magnet fields non-linear, and magnetic field distributions in such situations are very difficult to solve analytically. The most usual type of B-H dependence in charged particle optics is the one which describes the magnetic behavior of iron. In this case B rises monotonically with increasing H. Experimental data on B-H curves are widely available for many magnetic materials.

2.2   Boundary Conditions

The two basic integral equations of magnetostatics can be used to derive the boundary conditions of **B** and **H** at the interface between two regions, region 1 and region 2, whose relative permeabilities are different. It can be shown that

$$H_{t1} - H_{t2} = \frac{B_{t1}}{\mu_1} - \frac{B_{t2}}{\mu_2} = K \qquad B_{n1} = B_{n2} = \mu_1 H_{n1} = \mu_2 H_{n2} \qquad (1.49)$$

where H_{t1}, H_{n1}, B_{t1}, B_{n1} are the tangential and normal components of the magnetic intensity and flux density in region 1, H_{t2}, H_{n2}, B_{t2}, B_{n2} are the tangential and normal components of the magnetic intensity and flux density in region 2, and K is the current density at the interface. The relationship between the magnetic intensity and surface current density is better stated in vector form, since **H** is perpendicular to **K**, and they both lie in a plane which is perpendicular to the unit normal vector of the boundary **n**,

$$H_{t1} - H_{t2} = \mathbf{n} \times \mathbf{K} \qquad (1.50)$$

Where there are no surface currents, the tangential component of the magnetic intensity **H** will be continuous.

2.3   The Vector Potential

The point to point enforcement of the conservative nature of the electrostatic field means that the *curl* of the electric vector is everywhere zero

$$\oint \mathbf{E} \cdot \mathbf{dl} = 0 \Rightarrow \nabla \times \mathbf{E} = 0 \quad \forall x, y, z \tag{1.51}$$

This makes the electric scalar potential the most natural choice for electrostatic problems, since for any scalar function f(x,y,z), the following vector identity applies

$$\nabla \times \nabla f = 0 \tag{1.52}$$

The electric scalar potential u(x,y,z) springs from the conservative nature of the electric field. It simplifies Poisson's equation, in effect replacing a vector quantity (electric field) in the equation with a scalar one (potential). For a magnetic field, the situation is quite different. Here it is the divergence of **B** which is everywhere zero, and the equation which includes the source involves the *curl* of **B**, derived from Ampere's circuital law. This suggests the use of a different kind of potential. The following identity for any vector **A** is more appropriate

$$\nabla \cdot \nabla \times \mathbf{A} = 0 \Rightarrow \mathbf{B} = \nabla \times \mathbf{A} \tag{1.53}$$

Here **A** of course, refers to the well known magnetic vector potential. Although it guarantees that the divergence of **B** will always be zero, it makes the equation involving the magnetic source a more complicated one

$$\nabla \times \mathbf{H} = \mathbf{J} \Rightarrow \nabla \times \mu^{-1} \nabla \times \mathbf{A} = \mathbf{J} \tag{1.54}$$

This vector equation has three components, that is, it has three equations which are coupled together by the components of **A** (A_x, A_y, A_z). Each of these components are unknown functions of (x,y,z) and must be solved together with the boundary conditions.

It should be noted that the vector potential **A** is not unique. There is an extra degree of freedom that comes from $\nabla \times \nabla f = 0$, for any scalar function f, so

$$\mathbf{B} = \nabla \times (\mathbf{A} + \nabla f) = \nabla \times \mathbf{A} \tag{1.55}$$

It is usual to specify a gauge condition which in effect assumes that the function f(x,y,z) is constant

$$\nabla \cdot (\mathbf{A} + \nabla f) = \nabla \cdot \mathbf{A} = -\nabla^2 f = 0 \tag{1.56}$$

This added constraint on **A**, although easily specified analytically, is more difficult to apply when using numerical methods. Note that if it is only **B**(x,y,z) which is required, and not **A**(x,y,z), then the gauge condition need not be considered.

2.4   The Vector Potential in two dimensions

The main equation involving the source term **J** can be greatly simplified where only one component of the current exists and where there is some degree of symmetry.

One such case is for infinitely long sources where the current flows in one direction, as shown in Figure 1.6

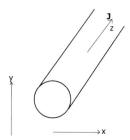

Figure 1.6 The Cartesian (x,y) plane

Here $\mathbf{J}$ has only the J_z component, and all derivatives with respect to z are assumed to be zero. Under these circumstances

$$\nabla \times \mathbf{H} = \frac{\partial H_y}{\partial x} - \frac{\partial H_x}{\partial y} = J_z \qquad (1.57)$$

Using the fact that in this case all derivatives with respect to z are zero, then

$$\mathbf{H} = \mu^{-1}\nabla \times \mathbf{A} \quad \Rightarrow \quad H_x = \mu^{-1}\frac{\partial A_z}{\partial y} \qquad H_y = -\mu^{-1}\frac{\partial A_z}{\partial x} \qquad (1.58)$$

This gives a scalar equation involving A_z

$$\mu^{-1}\left[\frac{\partial^2 A_z}{\partial x^2} + \frac{\partial^2 A_z}{\partial y^2}\right] = -J_z \qquad (1.59)$$

This equation is identical to Poisson's equation. In deriving this equation, it was assumed that the permeability μ is a constant, obviously this need not be the case. The more general formulation is

$$\frac{\partial}{\partial x}\left(\mu^{-1}\frac{\partial A_z}{\partial x}\right) + \frac{\partial}{\partial y}\left(\mu^{-1}\frac{\partial A_z}{\partial y}\right) = -J_z \qquad (1.60)$$

Note that the inherent symmetry for this situation ensures that the divergence of $\mathbf{A}$ is automatically zero, so no additional constraint on $\mathbf{A}$ is required. Once the vector potential distribution is found, the components of the magnetic flux density $B_x(x,y)$ and $B_y(x,y)$ are found by differentiating the vector potential with respect to x and y. The above equation together with boundary conditions on $A_z(x,y)$ are sufficient to guarantee zero divergence of the flux density.

Another important case is the rotationally symmetric magnetic layout, where the source current vector points in the azimuthal direction and axi-symmetric cylindrical coordinates (r,z) are used, as shown in Figure 1.7.

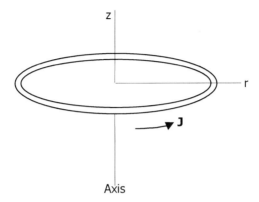

Figure 1.7 Cylindrical axisymmetry

Rotational symmetry means that all derivatives with respective to the azimuthal direction ϕ are zero, so that

$$\nabla \times \mathbf{H} = \frac{\partial H_r}{\partial z} - \frac{\partial H_z}{\partial r} = J_\phi \qquad (1.61)$$

Rotational symmetry also simplifies the relationship between **H** and **A**

$$\mathbf{H} = \mu^{-1} \nabla \times \mathbf{A} \quad \Rightarrow \quad H_r = -\mu^{-1} \frac{\partial A_\phi}{\partial z} \qquad H_z = \mu^{-1} \frac{1}{r} \frac{\partial (rA_\phi)}{\partial r} \qquad (1.62)$$

which leads to the following scalar equation

$$\frac{\partial}{\partial z}\left(\mu^{-1} \frac{\partial A_\phi}{\partial z} \right) + \frac{\partial}{\partial r}\left(\mu^{-1} \frac{1}{r} \frac{\partial (rA_\phi)}{\partial r} \right) = -J_\phi \qquad (1.63)$$

or

$$\frac{\partial}{\partial z}\left(\mu^{-1} \frac{\partial A_\phi}{\partial z} \right) + \frac{\partial}{\partial r}\left(\mu^{-1} \frac{\partial A_\phi}{\partial r} + \mu^{-1} \frac{A_\phi}{r} \right) = -J_\phi \qquad (1.64)$$

Here again, the divergence condition **A** is automatically satisfied due to symmetry. The above equation together with the boundary conditions on A_ϕ are sufficient to enforce the laws of magnetostatics point by point throughout the domain.

To find the components of the flux density, the spatial derivatives of the magnetic vector potential are required

$$B_r = -\frac{\partial A_\phi}{\partial z} \qquad B_z = \frac{1}{r}\frac{\partial}{\partial r}\left(rA_\phi\right) \tag{1.65}$$

While this might be a simple operation analytically, numerically, it can only be done to a certain precision. Where high accuracy is required, it is not a trivial task.

The above expression my be simplified in situations where $\mu(r,z)$ is constant to

$$\frac{\partial^2 A_\phi}{\partial z^2} + \frac{\partial^2 A_\phi}{\partial r^2} + \frac{1}{r}\frac{\partial A_\phi}{\partial r} - \frac{A_\phi}{r^2} = -\mu^{-1}J_\phi \tag{1.66}$$

There are two variations of this equation which are commonly used in electron optics. They are expressed in terms of the two functions, $P(r,z)$ and $F(r,z)$

$$P(r,z) = 2\frac{A_\phi(r,z)}{r} \qquad F(r,z) = 2\pi rA_\phi(r,z) \tag{1.67}$$

The components of **B** in terms of these functions are given by

$$B_z = \frac{\partial A_\phi}{\partial r} + \frac{A_\phi}{r} = P + \frac{r}{2}\frac{\partial P}{\partial r} = \frac{1}{2\pi r}\frac{\partial F}{\partial r} \tag{1.68}$$

$$B_r = -\frac{\partial A_\phi}{\partial z} = -\frac{r}{2}\frac{\partial P}{\partial z} = -\frac{1}{2\pi r}\frac{\partial F}{\partial z}$$

Integrating the expression for $B_z(r,z)$ from the axis (r=0) to a radius r=R, gives

$$F(R,z) = \int_0^R B_z(r,z)\cdot 2\pi rdr \tag{1.69}$$

This shows that $F(R,z)$ is the magnetic flux through a circular area around the axis of radius R in the plane where z=constant. Level lines of $F(r,z)$ will thus be identical to lines of magnetic flux. The function $P(r,z)$ on the axis (r=0) is equal to B_z

$$P(0,z) = B_z(0,z) \tag{1.70}$$

These expressions can simplify aberration calculations for an electron beam probe system.

2.5   The Magnetic Scalar Potential

Another way of formulating the magnetostatic problem is in terms of a scalar potential $V_m(x,y,z)$, in analogy to the electrostatic potential $u(x,y,z)$. This is only possible for regions where there is no current source

$$\nabla \times \mathbf{H} = 0 \Rightarrow \mathbf{H} = -\nabla V_m \qquad (1.71)$$

Now substituting this relation into the divergence of **B** yields Laplace's equation

$$\nabla \cdot \mu^{-1} \nabla V_m = 0 \qquad (1.72)$$

A scalar potential approach is sometimes advantageous, and traditionally, it was the first to be used in electron optics. But it has its own limitations and complications. Conceptually it might appear simple, since V_m has the units of amperes and can be directly related to the magneto-motive force of a magnetic circuit. The values of the scalar potential along a path from A to B are related to **H** by

$$\int_A^B \mathbf{H} \cdot \mathbf{dl} = -\left[V_m \right]_A^B \qquad (1.73)$$

It might seem that the source can be replaced by equivalent values of the scalar potential in a straightforward manner. But actually, the situation is more complicated. In fact V_m, unlike the electric potential, is a multi-valued function. This is because the magnetostatic field is not a conservative field. Moreover, restricting V_m to apply only to source-free regions does not make it independent of the source. In fact, the path A to B in the above integral may enclose a source, and still lie in a source free region. Consider for instance a circular path in the dielectric region of a coaxial cable. Upon completion of every cycle along such a path, the scalar magnetic potential jumps up discontinuously by I, the current flowing in the inner conductor. The only way of avoiding this is to make a branch cut in the path A to B, so that it does not enclose the inner conductor. Hence finding the boundary conditions for V_m in general is not a straightforward matter.

In certain situations, the scalar formulation can be used to calculate the fringe field distribution in the gap of a magnetic circuit. Consider the schematic diagram of a magnetic circuit shown in Figure 1.8. Here the region of interest lies around the air-gap, indicated by the dotted line.

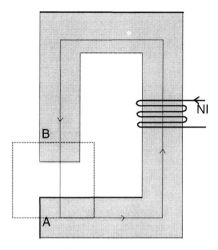

Figure 1.8 A magnetic circuit with an air-gap

When formulating the problem in terms of the magnetic scalar potential V_m, it is usually assumed that no drop of the magneto-motive force occurs in the iron between points A and B outside the fringe field region (outside the dotted box region), in which case

$$\int_A^B \mathbf{H} \cdot \mathbf{dl} = [V_m]_A - [V_m]_B \approx NI \qquad (1.74)$$

So the boundary conditions on the magnetic potential on the top and bottom region edges can be specified explicitly: -NI/2 below the gap (at point A), and NI/2 above the gap (at point B). The magnetic scalar distribution can then be solved in the region around the air gap. The boundary conditions at the other region edges can either be derived by linear interpolation, or specified to be Neumann boundaries. The magnetic flux density components are found by calculating the spatial derivatives of the magnetic scalar potential. The problem of finding the magnetic field distribution in the gap is now reduced to solving Laplace's equation (1.72). Note that the permeability $\mu(x,y,z)$ can vary from point to point.

2.6   The Reduced Scalar Potential

A more general scalar potential formulation, important to the solution of magnetostatic fields in three dimensions is based upon partitioning the magnetic intensity vector into a source part, $\mathbf{H_s}$, and a secondary magnetized part, $\mathbf{H_m}$, which is derived from the magnetic scalar potential $\phi(x,y,z)$

$$\mathbf{H} = \mathbf{H}_s + \mathbf{H}_m \tag{1.75}$$
$$\mathbf{H}_m = -\nabla\phi$$

If this expression is substituted into the divergence equation for $\mathbf{B}$, the following equation is obtained

$$\nabla\cdot\left(\mu\nabla\phi\right) = \nabla\cdot\left(\mu\mathbf{H}_s\right) = Q(x,y,z) \tag{1.76}$$

$Q(x,y,z)$ is an equivalent source term. This equation is in fact similar to Poisson's equation since μ is a known function of x,y and z, and $\mathbf{H}_s(x,y,z)$ can be found from Biot-Savart's law, which is in effect, the point to point solution of Ampere's circuital law. This is because $\nabla\times(\nabla\phi)$ is zero, so

$$\nabla\times\left(\mathbf{H}_s - \nabla\phi\right) = \nabla\times\mathbf{H}_s = \mathbf{J} \tag{1.77}$$

and

$$\mathbf{dH}_s = \frac{1}{4\pi}\frac{\mathrm{NIdL}\times\mathbf{r}}{|\mathbf{r}|^3}$$

where $\mathbf{r}$ is the vector from a point on the source to the field point, NI is the source current vector, and dL is an incremental distance along the source path, as shown in Figure 1.9a.

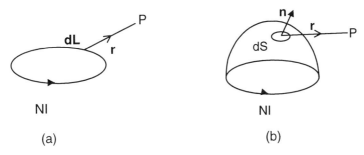

Figure 1.9 Field from current carrying conductor

Obviously $\mathbf{H}_s(x,y,z)$ can be found from integrating over the whole source region. In most cases, this amounts to integrating along wire loops of current. Once this has been done, the equivalent source function $Q(x,y,z)$ can be calculated. Note that for linear systems where iron regions have constant permeability, $Q(x,y,z)$ will only be non-zero at iron/air interfaces. For the more general case where iron regions are saturated, $Q(x,y,z)$ will be non-zero inside the iron. After the calculation of the equivalent source function, a Poisson-like equation in terms of the reduced potential $\phi(x,y,z)$ needs to be solved.

Although the above formulation, known as the 'reduced scalar potential' method, is more general than the previous one, its numerical implementation still has problems.

One such problem is that of field cancellation. In many situations $\mathbf{H}_s$ and $\mathbf{H}_m$ are of approximately the same magnitude but opposite in direction, so cancellation occurs in computing the total $\mathbf{H}$, giving a loss of accuracy. This error is particularly large for $\mathbf{B}$ in regions of high permeability, since the cancellation effect is amplified by μ,

$$\mathbf{B} = \mu\left(\mathbf{H}_s - \nabla\phi\right) \tag{1.78}$$

In situations where there is no iron material, a scalar potential distribution can be obtained from a variant of Bio-Savart's law

$$\phi(x, y, z) = -\frac{NI}{4\pi} \iint_S \frac{\mathbf{r} \times \mathbf{n} \, dS}{|\mathbf{r}|^3} \tag{1.79}$$

The surface S, and the normal vector $\mathbf{n}$ from a small part of it dS is illustrated in Figure 1.9b. This expression for the scalar potential is useful in calculating the source field distribution for magnetic deflectors. Note that the surface S is topologically different to the shape of the coils. In the case of magnetic deflectors, while the coil can only be described in three dimensions, the surfaces between the coils have symmetry, allowing for the use of two dimensional scalar potentials.

2.7   The Two Scalar Potential approach

To avoid the cancellation errors in the reduced scalar potential approach, a total scalar potential ψ can be used to model iron regions. We have

$$\mathbf{H} = -\nabla\psi \tag{1.80}$$

And since this describes the total magnetic field in the iron region, ψ is called the total scalar potential. However, the total potential approach is possible only for regions free of current sources, and it cannot be applied to current source regions. Fortunately in practice, the whole domain Ω can be divided into two regions Ω_j and Ω_k, which are usually separate. Ω_k contains iron regions, while Ω_j encloses both the air and current regions. Let the interface between the two regions be denoted by Γ_{jk}. These regions are shown in Figure 10.

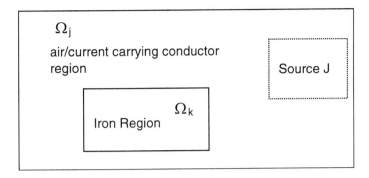

Figure 1.10 Region Ω showing the conductor, iron and air regions

In Ω_j, the reduced scalar potential can still be used, as μ is low (usually assumed to be μ_0). In Ω_k, the total scalar potential ψ is used instead.   Hence in the source region Ω_j, we have

$$\mathbf{H} = \mathbf{H_s} - \nabla\phi \tag{1.81}$$

Giving

$$\nabla\cdot\mu\nabla\phi = \nabla\cdot\mu\mathbf{H_s} \tag{1.82}$$

In Ω_k

$$\mathbf{H} = -\nabla\psi \tag{1.83}$$

giving

$$\nabla\cdot\mu\nabla\psi = 0 \tag{1.84}$$

The two above equations must satisfy the usual boundary conditions at the region interface. The first is to ensure the continuity of the B-field normal component

$$\mu(-\nabla\psi)\cdot\mathbf{n} = \mu_0\,(\mathbf{H_s} - \nabla\phi)\cdot\mathbf{n} \tag{1.85}$$

or

$$-\mu\frac{\partial\psi}{\partial n} = \mu_0\left(H_{sn} - \frac{\partial\phi}{\partial n}\right) \tag{1.86}$$

where $\mathbf{n}$ is the outward normal flux from Ω_k, and H_{sn} is the source field normal component.

The second boundary condition guarantees the continuity of the tangential component of **H**

$$(-\nabla\psi)\cdot\mathbf{t} = (\mathbf{H_s} - \nabla\phi)\cdot\mathbf{t} \qquad\qquad (1.87)$$

or

$$-\frac{\partial\psi}{\partial t} = \mathbf{H}_{st} - \frac{\partial\phi}{\partial t} \qquad\qquad (1.88)$$

where **t** is the tangent direction along the interface Γ_{jk}, and H_{st} is the source field tangential component.

Integrating the above equation over a path along Γ_{jk} gives an integral relationship between the two scalar potentials at the interface

$$\psi = \phi - \int_{\Gamma_{jk}} \mathbf{H_s} \cdot \mathbf{t} ds = \phi - \int_{\Gamma_{jk}} H_{st} dS \qquad\qquad (1.89)$$

where $\mathbf{H_s}$ is obtained from Bio-Savart's Law. Note that in this case, $\mathbf{H_s}$ needs to be evaluated only at the iron/air interface, even when saturation occurs.

2.8   The Vector Potential in three dimensions

Consider once again the fundamental magnetostatics vector potential equation

$$\nabla\times\mu^{-1}\nabla\times\mathbf{A} = \mathbf{J} \qquad\qquad (1.90)$$

The three components of this equation can either be solved directly, or a gauge condition can be used. The solution of this vector equation is found in terms of the three component distributions, **A** (A_x, A_y, A_z). To simplify the discussion, assume that the relative permeability is constant. The above equation can then be expanded to give

$$\nabla(\nabla\cdot\mathbf{A}) - \nabla^2\mathbf{A} = \mathbf{J}\mu \qquad\qquad (1.91)$$

If the gauge condition of zero divergence on **A** is used, the solution to the magnetostatic problem must involve solving the following pair of coupled equations

$$\begin{aligned}\nabla^2\mathbf{A} &= \mathbf{J}\mu \\ \nabla(\nabla\cdot\mathbf{A}) &= 0\end{aligned} \qquad\qquad (1.92)$$

Note that the first of these equations is a vector equation and has three components, so altogether, there are four equations to solve. While these equations are simply stated in their general form, they are difficult to solve, both analytically and numerically.

2.9   Magnetic Energy

The energy W of a magnetic field is usually expressed in the following way

$$dW = \int_V \mathbf{H} \cdot \mathbf{dB} dV \tag{1.93}$$

Where V is the volume of the domain. This expression accounts for saturation, where **H** is function of B, **H(B)**. For linear fields where **H** is proportional to **B**, this simplifies to

$$W = \frac{1}{2} \int_V \mathbf{H} \cdot \mathbf{B} dV \tag{1.94}$$

Obviously for either the linear or saturated case, the magnetic energy of the system can be formulated in terms of a scalar or vector potential.

Chapter 2
FIELD SOLUTIONS FOR CHARGED PARTICLE OPTICS

This chapter gives an introduction to how field distributions are used in charged particle optics, and is primarily intended for the non-specialist reader. There are of course many books on this subject and the reader is encouraged to consult them for more details [1]. The simulation of charged particle beam systems involves the plotting of trajectory paths through electric and magnetic fields. In most cases, charged particle beam systems can be adequately modelled by assuming these fields are static or quasi-static. In instruments like high energy particle accelerators, or RF tubes, the interaction of charged particles with electromagnetic fields must be considered, but this situation will not be treated here. Jianming Jin gives a good introduction to the finite element for electromagnetic field problems [2].

1.   THE EQUATIONS OF MOTION

The basic equation governing the motion of charged particles through electric and magnetic fields is the Lorentz force vector equation

$$\mathbf{F} = q(\mathbf{E} + \mathbf{v} \times \mathbf{B})\tag{2.1}$$

where $\mathbf{F}$ is the force on the charged particle, q is the charge, $\mathbf{E}$ is the electric field, $\mathbf{v}$ is the particle's velocity, and $\mathbf{B}$ is the magnetic field. For simplicity, relativistic effects have been neglected. In rectilinear coordinates (x,y,z), this equation decouples into three separate equations

$$m\ddot{x} = q\left[E_x + \left(v_y B_z - v_z B_y\right)\right]$$
$$m\ddot{y} = q\left[E_y + \left(v_z B_x - v_x B_z\right)\right]\tag{2.2}$$
$$m\ddot{z} = q\left[E_z + \left(v_x B_y - v_y B_x\right)\right]$$

where E_x, E_y, E_z, B_x, B_y, and B_z are the electric and magnetic field components in the x, y and z directions.

Spatial variation of the field components usually makes the above set of equations non-linear, and so it is usual to solve them numerically. Note that the initial conditions, that is the particle's starting position and velocity, as well as the field distribution at every point in the domain must be known in advance. Since the field distribution is usually formulated in terms of either the scalar or vector potential, then in most cases, numerical differentiation is also required, and this may turn out to be a factor limiting the final accuracy to which the particle's trajectory can be plot.

The above equations of motion are used when direct ray tracing of a particle's motion is required. This occurs often in charged particle optics. It is required in the simulation of electron guns, as shown in Figure 2.1, where electron trajectories are traced from the cathode of a field emission electron gun.

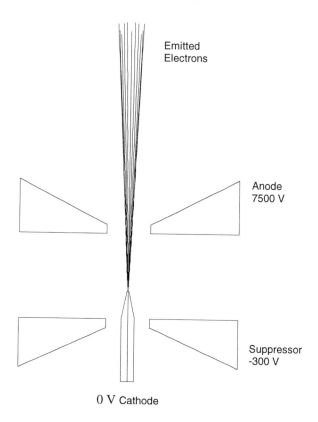

Figure 2.1 Electron trajectories in a field emission electron gun.

Direct ray tracing is also needed for the collection or energy filtering of secondary electrons. Figure 2.2 shows a schematic drawing of the emitted secondary electrons from the specimen to a scintillator detector in a Scanning Electron Microscope. The primary beam of electrons strike the specimen at energies that range from 1 keV to 30 keV, and the greater proportion of the secondary electrons have energies below a few eV. In all these types of situations, the spatial derivative of the electric or magnetic potential must be accurately known at any point in the region where electrons are in motion. This is a special requirement of charged particle optics that does not usually exist for other applications. Normally, it is sufficient to have potential or field information at the nodes of the numerical mesh, and even then, the precision to which they are required is not usually critical.

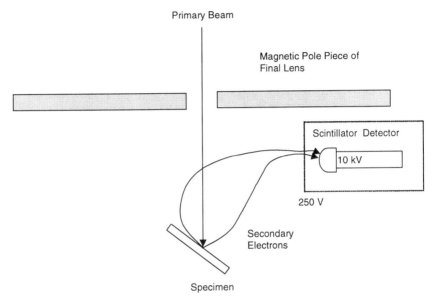

Figure 2.2 Secondary electrons in a Scanning Electron Microscope

There are also other ways in which field distributions for charged particle optics differ from the field distributions found in other applications. The primary beam of charged particle probe systems is usually confined to a small area around an optic axis. In most cases, the optic axis lies at the centre of an axisymmetric cylindrical coordinate system, one which is defined in terms of the off-axis radius r and the longitudinal coordinate z, (r,z). Even where the optic axis is curved, a local axisymmetric coordinate system is used, as shown in Figure 2.3.

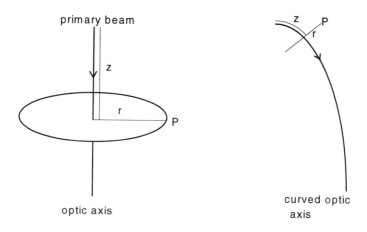

Figure 2.3 Electron Beam Optic Axis

Figure 2.4 shows an example of a magnetic objective lens design for a Scanning Electron Microscope.

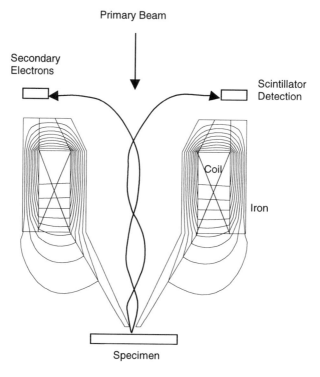

Figure 2.4 A Scanning Electron Microscope magnetic objective lens design

The magnetic or electric fields which focus the primary beam are usually of the round lens type, which means that they are axisymmetric field distributions defined in terms of (r,z): the scalar potential $U(r,z)$ for electric fields and the vector potential component $A_\phi(r,z)$ for the magnetic field. In cylindrical coordinates, (r,ϕ,z), the equation of motion for an electron simplifies to the following set of coupled equations

$$\frac{d^2r}{dt^2} - r(\dot{\phi})^2 = \frac{e}{m}\left[\frac{\partial U}{\partial r} - \dot{\phi}\frac{\partial(rA_\phi)}{\partial r}\right]$$

$$\frac{d(r^2\dot{\phi})}{dt} = \frac{e}{m}\frac{d(rA_\phi)}{dt} \qquad\qquad (2.3)$$

$$\frac{d^2z}{dt^2} = \frac{e}{m}\left[\frac{\partial U}{\partial z} - r\dot{\phi}\frac{\partial A_\phi}{\partial z}\right]$$

where e is the electronic charge of an electron. By integrating the middle equation, these equations can be further simplified to

$$\frac{d^2r}{dt^2} = \frac{e}{m}\frac{\partial \Phi}{\partial r}$$

$$\frac{d^2z}{dt^2} = \frac{e}{m}\frac{\partial \Phi}{\partial z} \qquad\qquad (2.4)$$

$$\frac{d\phi}{dt} = \frac{e}{m}\left(\frac{A_\phi}{r} + \frac{c}{r^2}\right)$$

where c is an integration constant related to the initial angular velocity, and

$$\Phi(r,z) = U(r,z) - \frac{e}{2m}\left[A_\phi(r,z) + \frac{c}{r}\right]^2 \qquad\qquad (2.5)$$

The potential $\Phi(r,z)$ is known as Störmer's potential, and requires knowledge of the particle's initial angular velocity, the electric potential and the magnetic vector potential. These equations of motion can also be used for direct ray tracing. Note that once again, it is the derivative of the potential distribution that enters the equations of motion, which will again necessitate the use of numerical differentiation.

2.   THE PARAXIAL EQUATION OF MOTION

Since the primary beam travels very close to the optic axis, the off-axis field distributions can be expressed in terms of the axis field distributions. For a small region around the optic axis, consider Laplace's equation in axisymmetric coordinates for the electrostatic potential $U(r,z)$

$$\frac{\partial^2 U}{\partial z^2} + \frac{\partial^2 U}{\partial r^2} + \frac{1}{r}\frac{\partial U}{\partial r} = 0 \qquad (2.6)$$

Let the solution to the above equation be written as a power series in r

$$U(r, z) = a_0(z) + a_2(z)r^2 + a_4(z)r^4 + ... \qquad (2.7)$$

This function is an even function of r, due to the inherent symmetry about the optic axis, $U(r,z)=U(-r,z)$. If we assume that $a_0(z)=U(0,z)=U_0(z)$, it is possible to calculate the above coefficients by substituting the expansion for $U(r,z)$ into Laplace's equation, so that

$$U(r, z) = U_0(z) - \frac{r^2}{4}\frac{\partial^2 U_0(z)}{\partial z^2} + \frac{r^4}{64}\frac{\partial^4 U_0(z)}{\partial z^4} - \frac{r^6}{2304}\frac{\partial^6 U_0(z)}{\partial z^6} + ... \qquad (2.8)$$

This expansion only involves knowledge of $U_0(z)$ and its derivatives. Obviously the near axis electric field components can be found from differentiating the above expression by r or z.

For the near axis flux density components $B_z(r,z)$ and $B_r(r,z)$, the starting point is the axisymmetric field equation involving the vector potential component A_ϕ

$$\frac{\partial^2 A_\phi}{\partial z^2} + \frac{\partial^2 A_\phi}{\partial r^2} + \frac{\partial}{\partial r}\left(\frac{A_\phi}{r}\right) = 0 \qquad (2.9)$$

The right-hand side of this equation is zero since the near axis-region is invariably a source-free region. In this case, an odd expansion in A_ϕ is used since it is zero on the optic axis

$$A_\phi(r, z) = c_1(z)r + c_3(z)r^3 + c_5(z)r^5 + ... \qquad (2.10)$$

Following the same procedure as before, substituting the polynomial expansion for $A_\phi(r,z)$ in the above magnetostatic equation, the following expansion of $A_\phi(r,z)$ is obtained

$$A_\phi(r, z) = c_1(z)r - \frac{1}{8}\frac{\partial^2 c_1(z)}{\partial z^2}r^3 + \frac{1}{192}\frac{\partial^4 c_1(z)}{\partial z^4}r^5 + ... \qquad (2.11)$$

The components $B_z(r,z)$ and $B_r(r,z)$ around the optic axis are easily found in terms of the above expression

$$B_z(r, z) = \frac{\partial A_\phi}{\partial r} + \frac{A_\phi}{r} = B_0(z) - \frac{1}{4}\frac{\partial^2 B_0(z)}{\partial z^2}r^2 + \frac{1}{64}\frac{\partial^4 B_0(z)}{\partial z^4}r^4 ....$$

$$B_r(r, z) = -\frac{\partial A_\phi}{\partial z} = -\frac{1}{2}\frac{\partial B_0(z)}{\partial z}r + \frac{1}{16}\frac{\partial^3 B_0(z)}{\partial z^3}r^3 .... \qquad (2.12)$$

where $B_0(z) = 2c_1(z)$

Clearly from the above expressions, high-order derivatives of the electric and magnetic potentials along the optic axis are required for points far off-axis.

Here, it is assumed that the primary beam is travelling close to the optic axis, which is valid for instruments such as electron microscopes. To determine the first-order optical properties, such as the beam's focal position or focal length, it is sufficient to make the following approximations:

$$U(r,z) \approx U_0(z) - \frac{1}{4} \frac{\partial^2 U_0(z)}{\partial z^2}$$

$$E_z(r,z) \approx -\frac{\partial U_0(z)}{\partial z}$$

$$E_r(r,z) \approx -\frac{r}{2} \frac{\partial^2 U_0(z)}{\partial z^2}$$

$$A_\phi(r,z) \approx \frac{r}{2} B_0(z)$$

$$B_z(r,z) \approx B_0(z)$$

$$B_r(r,z) \approx -\frac{r}{2} \frac{\partial B_0}{\partial z}$$

$$(2.13)$$

Another approximation that can be made when considering charged particle probe systems is that the longitudinal velocity is much greater than the transverse velocity. The total velocity v is then related to the axial potential distribution by $U_0(z)$

$$v^2 \approx \left(\frac{dz}{dt}\right)^2 = \frac{2eU_0(z)}{m} \qquad (2.14)$$

assuming that the initial velocity and the potential at the emission point are both zero. The foregoing approximations lead to the well known 'paraxial' equation

$$\frac{d^2r}{dz^2} + \frac{1}{2U_0} \frac{dU_0}{dz} \frac{dr}{dz} + \left(\frac{1}{4U_0} \frac{d^2U_0}{dz^2} + \frac{e}{8mU_0} B_0^2\right) r = 0 \qquad (2.15)$$

This is a second order equation for $r(z)$ that can be solved numerically. It requires the first and second order derivatives of the axial potential $U_0(z)$, and the magnetic axial field distribution $B_0(z)$.

The paraxial equation has two independent solutions, $r_1(z)$ and $r_2(z)$, so that the general solution $r(z)$ is given by

$$r(z) = Cr_1(z) + Dr_2(z) \qquad (2.16)$$

where C and D are constants which can be solved by specifying the initial conditions for the ray, typically using the initial value of r and its slope dr/dz. Two independent ray paths are shown in Figure 2.5. Normally the slope of $r_2(z)$ and the starting point of $r_1(z)$ are specified to be zero. The point at which a paraxial ray focuses is called the Gaussian plane. Here the Gaussian plane is located at the focal point for the ray $r_1(z)$ and the linear magnification of the lens M can be found from the value of ray $r_2(z)$ on the Gaussian plane.

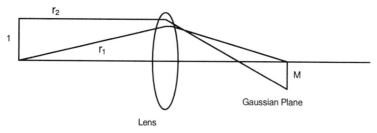

Figure 2.5 Two independent ray solutions

3.   ON-AXIS LENS ABERRATIONS

From considerations of first-order optics alone, the final predicted spot size for the primary beam would be zero, but in practice, the final probe size is limited by lens aberrations. Lens aberrations have been discussed in many books and only a brief summary of them will be given here for the simple situation of an undeflected primary beam. The purpose of discussing them here is to illustrate the important role played by axial field distributions and their derivatives in the subject of computational charged particle beam optics.

The first class of aberrations is geometrical in nature, the simplest of which is known as spherical aberration. This effect arises from the fact that the focal length of a lens is dependent on the off-axis distance with which a charged particle enters the lens field, as shown in Figure 2.6.

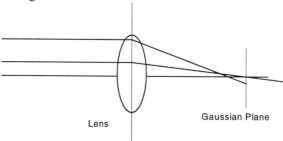

Figure 2.6 The Spherical aberration effect

The broadening of the probe size Δr_s at the Gaussian plane due to spherical aberration is often expressed in the following way

$$\Delta r_s = C_s \alpha^3 \qquad (2.17)$$

where α is the semi-angle at the image plane, and C_s is called the coefficient of spherical aberration.

Another aberration comes from the energy dispersion in the charged particle beam and is known as chromatic aberration. A ray at energy eU, focuses at a different point to a ray having an energy e(U+ΔU), as illustrated in Figure 2.7.

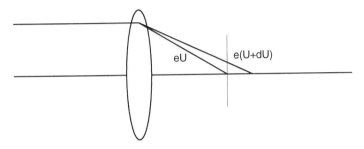

Figure 2.7 The Chromatic aberration effect

The contribution of chromatic aberration to the final spot size Δr_c is defined by

$$\Delta r_c = C_c \alpha \frac{\Delta U}{U} \qquad (2.18)$$

where U is the beam energy at the point of focus, ΔU is the beam voltage spread at the source (gun), and C_c is called the coefficient of chromatic aberration.

Another contribution to the primary beam final spot size is due to the wave-like nature of the charged particle, and is known as diffraction aberration. This type of aberration is usually treated as field independent, and hence will not be discussed in detail here. It is dependent on the beam voltage U and the final semi-angle α. Diffraction broadens the final spot by Δr_d and for an electron beam it is given by

$$\Delta r_d = \frac{0.735 \times 10^{-9}}{\alpha \sqrt{U}} \qquad (2.19)$$

The total aberration limited probe spot size for an undeflected   primary beam is usually found by adding each aberration term in quadrature

$$\Delta r^2 = \left(\frac{\Delta r_s}{4}\right)^2 + \left(\frac{\Delta r_c}{2}\right)^2 + \Delta r_d^2 \qquad (2.20)$$

The extra factors of reduction in the contributions from the spherical and chromatic aberration terms are due to moving the image plane to a point where the spot size is minimised, and is known as the 'plane of least confusion'.

A perturbation method on the paraxial equation is used to derive C_s and C_c. In the case of spherical aberration, higher order contributions in the field expansions are retained, while the derivation of the chromatic aberration coefficient proceeds by perturbation on the beam voltage in the paraxial equation. To illustrate this procedure, consider the case of a purely electrostatic lens where the expansion for the potential $U(r,z)$ includes an extra higher-order term

$$U(r,z) = U_0(z) - \frac{r^2}{4}\frac{\partial^2 U_0(z)}{\partial z^2} + \frac{r^4}{64}\frac{\partial^4 U_0(z)}{\partial z^4} \tag{2.21}$$

Substituting into the equations of motions in cylindrical coordinates

$$\frac{d^2 r}{dt^2} = \frac{e}{m}\frac{\partial U}{\partial r} = \frac{e}{m}\left(-\frac{r}{2}\frac{\partial^2 U_0}{\partial z^2} + \frac{r^3}{16}\frac{\partial^4 U_0}{\partial z^4}\right)$$

$$\frac{d^2 z}{dt^2} = \frac{e}{m}\frac{\partial U}{\partial z} = \frac{e}{m}\left(\frac{\partial U_0}{\partial z} - \frac{1}{4}\frac{\partial^3 U_0}{\partial z^3}r^2\right) \tag{2.22}$$

Note that in this case, there is no rotation in the azimuthal case since the magnetic field is assumed to be zero. The double derivatives with respect to time can be eliminated from these equations by using the first-order approximation to the energy equation given by (2.14), so

$$\frac{d^2 r}{dz^2}(\dot{z})^2 + \frac{e}{m}\frac{\partial U}{\partial z}\frac{dr}{dz} = \frac{e}{m}\frac{\partial U}{\partial r} \tag{2.23}$$

The single derivative with respect to time can be eliminated by using the following higher order approximation to the energy equation

$$(\dot{z})^2 \approx \frac{2eU_0}{m}\left(1 - \frac{1}{4U_0}\frac{\partial^2 U_0}{\partial z^2}\right)\left(1 - \left(\frac{dr}{dz}\right)^2\right) \tag{2.24}$$

If $r(z)=r_p(z)+\varepsilon(z)$, where $r_p(z)$ is the paraxial solution, and $\varepsilon(z)$ is a small perturbation which arises from the higher-order terms in the field expansion, then it can be shown that by ignoring terms in r^4 and above, the following equation of motion is obtained

$$\frac{d^2\varepsilon}{dz^2} + \frac{1}{2U_0}\left(\frac{\partial U_0}{\partial z}\right)\left(\frac{d\varepsilon}{dz}\right) + \frac{1}{4U_0}\frac{\partial^2 U_0}{\partial z^2}\varepsilon = s(z)$$

where

$$s(z) = \left[\frac{1}{32U_0} \frac{\partial^4 U_0}{\partial z^4} - \frac{1}{16U_0^2} \left(\frac{\partial^2 U_0}{\partial z^2} \right)^2 \right] r_p^3$$

$$+ \frac{1}{8} \left[\frac{1}{U_0} \frac{\partial^3 U_0}{\partial z^3} - \frac{1}{U_0^2} \frac{\partial U_0}{\partial z} \frac{\partial^2 U_0}{\partial z^2} \right] r_p^2 \frac{dr_p}{dz}$$

$$- \frac{1}{4U_0} \frac{\partial^2 U_0}{\partial z^2} \left(\frac{dr_p}{dz} \right)^2 r_p - \frac{1}{2U_0} \frac{\partial U_0}{\partial z} \left(\frac{dr_p}{dz} \right)^3$$

(2.25)

The above equation is a second order differential equation in $\varepsilon(z)$ with respect to z, and can be solved by the use of the Variation of Parameters method which involves using two independent solutions $r_1(z)$ and $r_2(z)$. The value of $\varepsilon(z)$ at the Gaussian (image) plane z_i is defined in terms of the coefficient of spherical aberration

$$\varepsilon(z_i) = C_s \alpha^3 \tag{2.26}$$

After some simplification, C_s is found to be

$$C_s = \frac{M}{16\sqrt{U_0}(z_0)} \int_{z_0}^{z_i} r_1(z)^4 \sqrt{U_0(z)} \left[f + 4 \frac{g}{r_1(z)} \frac{dr_1(z)}{dz} + 2 \frac{h}{r_1(z)^2} \left(\frac{dr_1(z)}{dz} \right)^2 \right] dz$$

where

$$f = \frac{5}{4U_0^2} \left(\frac{\partial^2 U_0}{\partial z^2} \right)^2 + \frac{5}{24U_0^4} \left(\frac{\partial U_0}{\partial z} \right)^4$$

$$g = \frac{7}{6U_0^3} \left(\frac{\partial U_0}{\partial z} \right)^3 \tag{2.27}$$

$$h = -\frac{3}{4U_0^2} \left(\frac{\partial U_0}{\partial z} \right)^2$$

Here z_0 is the value of z at the object plane, and M represents the linear magnification. For a detailed derivation of the above expression, the reader is referred to Grivet [3]. The integral for C_s is usually integrated numerically.

The spherical aberration depends on the paraxial ray solution as well as the axial electric potential distribution $U_0(z)$. Note that it also requires knowledge of the first and second order derivatives of $U_0(z)$.

In situations where a magnetic field exists, the following high-order terms can be included

$$B_z(r, z) \approx B_0(z) - \frac{1}{4} \frac{\partial^2 B_0(z)}{\partial z^2}$$

$$B_r(r, z) \approx -\frac{1}{2} \frac{\partial B_0(z)}{\partial z} r + \frac{1}{16} \frac{\partial^3 B_0(z)}{\partial z^3} r^3 \tag{2.28}$$

Putting these expressions into the equations of motion, and ignoring terms in r^4 and above, the expressions for the functions f, g, h in equation (2.27) are modified to

$$f = \frac{5}{4U_0^2} \left(\frac{\partial^2 U_0}{\partial z^2} \right)^2 + \frac{5}{24U_0^4} \left(\frac{\partial U_0}{\partial z} \right)^4 + \frac{e}{mU_0} \left(\frac{\partial B_0}{\partial z} \right)^2 + \frac{3}{8} \left(\frac{e}{mU_0} \right)^2 B_0^2$$

$$+ \frac{35}{16} \frac{e}{mU_0^2} \left(\frac{\partial U_0}{\partial z} \right)^2 B_0^2 - \frac{3e}{mU_0^2} \frac{\partial U_0}{\partial z} \frac{\partial B_0}{\partial z} B_0 \tag{2.29}$$

$$g = \frac{7}{6U_0^3} \left(\frac{\partial U_0}{\partial z} \right)^3 - \frac{e}{2mU_0^2} \frac{\partial U_0}{\partial z} B_0^2$$

$$h = -\frac{3}{4U_0^2} \left(\frac{\partial U_0}{\partial z} \right)^2 - \frac{e}{2mU_0} B_0^2$$

The above expressions require knowledge of $B_0(z)$ and its first derivative, which are usually obtained from finite element solved potential distributions.

The derivation for the coefficient of chromatic aberration C_c starts with the paraxial equation

$$\frac{d^2r}{dz^2} + \frac{1}{2U_0} \frac{dU_0}{dz} \frac{dr}{dz} + \left(\frac{1}{4U_0} \frac{d^2U_0}{dz^2} + \frac{e}{8mU_0} B_0^2 \right) r = 0 \tag{2.30}$$

A perturbation in the beam voltage ΔU is assumed to give rise to a change in the paraxial ray solution by $\varepsilon(z)$. The paraxial equation is now changed to

$$\frac{d^2r}{dz^2} + \frac{1}{2(U_0 + \Delta U)} \frac{dU_0}{dz} \frac{dr}{dz} + \left(\frac{1}{4(U_0 + \Delta U)} \frac{d^2U_0}{dz^2} + \frac{e}{8m(U_0 + \Delta U)} B_0^2 \right) r = 0 \tag{2.31}$$

Assuming $\Delta U/U_0$ is small, typically less than 0.001, and putting, $r(z) = r_p(z) + \varepsilon(z)$, then the above equation can be simplified to

$$\frac{d^2\varepsilon}{dz^2} + \frac{1}{2U_0} \frac{dU_0}{dz} \frac{d\varepsilon}{dz} + \left(\frac{1}{4U_0} \frac{d^2U_0}{dz^2} + \frac{e}{8mU_0} B_0^2 \right) \varepsilon = s(z)$$

where

$$s(z) = \Delta U \left[\frac{1}{2U_0^2} \frac{dU_0}{dz} \frac{dr_p}{dz} + \frac{1}{4U_0^2} \frac{d^2U_0}{dz^2} r_p + \frac{e}{8mU_0^2} r_p B_0^2 \right] \tag{2.32}$$

Again, the Variation of Parameters technique can be used, involving two independent solutions $r_1(z)$ and $r_2(z)$. The value of $\varepsilon(z)$ at the Gaussian (image) plane z_i is defined in terms of the coefficient of chromatic aberration by

$$\varepsilon(z_i) = \frac{\Delta U}{U_0(z_i)} C_s \alpha \qquad (2.33)$$

After some simplification, C_c is derived to be

$$C_c = M\sqrt{U_0(z_0)} \int_{z_0}^{z_i} \frac{r_1(z)}{\sqrt{U_0}} \left[\frac{1}{2U_0} \frac{dU_0}{dz} \frac{dr_1(z)}{dz} + \frac{1}{4U_0} \frac{d^2U_0}{dz^2} r_1(z) + \frac{e}{8mU_0} r_1(z)B_0^{\,2} \right]$$

$$(2.34)$$

4.   ELECTROSTATIC AND MAGNETIC DEFLECTION FIELDS

4.1   The Deflector Layout

Magnetic deflection systems usually have the current-carrying conductors arranged on a surface of revolution about the z-axis in the form of a cylinder or a tapered horn. The windings may be in the form of either saddle or toroidal type, and the deflector itself can be placed next to rotationally symmetric magnetic materials. Toroidal and saddle deflection yokes are shown in Figure 2.8. The arrows indicate the direction of current flow. In this figure, the current direction in the coils is in anti-symmetry with respect to the z-axis. The resulting field will be mainly in the x-direction, and they will be referred to here as x-deflectors or x-coils. The y-coils (not shown in the figure) are of identical design, and are merely rotated 90° relative to the x-coils.

A typical electrostatic deflector is shown in Figure 2.9. The number of electrodes is usually a multiple of four, the most common arrangement being either a quadrupole or an octopole deflector. Here the applied voltages on the electrodes have symmetry with respect to x and anti-symmetry with respect to y, therefore the field will be mainly in the x-direction and therefore they can be regarded as x-electrodes. Similar to y-coils, y-electrodes are obtained by rotating the x-electrodes by 90°. Most deflectors have fourfold symmetry with respect to the optic axis and this forms the basis of the harmonic analysis for the deflection field.

All deflection fields considered here are assumed to be static or quasic-static. For a magnetic deflector, $\mathbf{H}$ is partitioned into a source field $\mathbf{F}$ and a secondary magnetised part related to the scalar potential Φ

$$\mathbf{H} = \nabla\Phi + \mathbf{F} \qquad (2.35)$$

F is a vector function related to the excitation current. In physical terms, the scalar potential Φ suffices to represent the field everywhere except at the coil windings, while the vector function is used to take account of the jump in the field **H** which occurs at the coil windings.

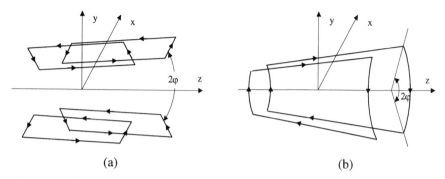

Figure 2.8 Typical magnetic deflection yokes; (a) toroidal coil; and (b) saddle coil. (Only x-deflectors are shown, arrows indicate the direction of current flow).

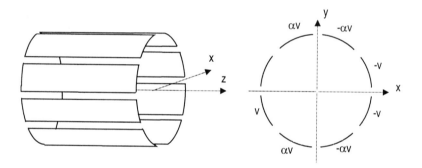

Figure 2.9 A typical electrostatic deflector; (a) perspective view, (b) end view
(The electrode potentials shown are corresponding to x-deflector)

Substituting this expression for **H** in the two basic magnetostatic equations (1.46) and (1.48) and invoking the fact that $\nabla \times (\nabla \Phi) = 0$ for any scalar function Φ, this gives

$$\nabla \times \mathbf{F} = \mathbf{J} \tag{2.36}$$

$$\nabla \cdot [\mu(\nabla \Phi + \mathbf{F})] = 0 \tag{2.37}$$

where J is the source current. The vector function **F** related to the source can be defined as being equal to zero everywhere outside the winding, and has the direction of the surface normal to the area enclosed by the conductors.

For the electrostatic deflection field in space-charge free regions, the scalar electric potential Φ obeys Laplace's equation

$$\nabla \cdot [\varepsilon(\nabla\Phi)] = 0 \qquad (2.38)$$

Since the electrostatic and magnetic deflection fields are always calculated separately, we can use Φ to represent both the magnetic scalar potential and the electric potential. In this case, equations (2.37) and (2.38) can thus be combined as

$$\nabla \cdot [\gamma(\nabla\Phi + k\mathbf{F})] = 0 \qquad (2.39)$$

where $\gamma=\mu$, $k=1$ for magnetic deflection and $\gamma=\varepsilon$, $k=0$ for electrostatic deflection.

4.2   Harmonic expansions of the scalar potential

On account of the fourfold symmetry of the x-deflector (coils or electrodes), the electric potential or magnetic scalar potential, generally represented by Φ, has the following symmetry properties

$$\left.\begin{array}{l} \Phi(r,z,-\theta) = \Phi(r,z,\theta) \\[2ex] \Phi(r,z,\theta+\pi) = -\Phi(r,z,\theta) \end{array}\right\} \qquad (2.40)$$

Consequently, if Φ is expressed as a Fourier series of harmonic components, the series will contain only cosine terms of odd order, and can therefore be expressed in the general form

$$\Phi(r,z,\theta) = \sum_{m=1,3,5...}^{\infty} \Phi_m(r,z,\theta)\cos m\theta \qquad (2.41)$$

Since Φ obeys Laplace's equation, it follows that function $\Phi_m(r,z)$ obeys the equation

$$\frac{\partial^2\Phi_m}{\partial r^2} + \frac{1}{r}\frac{\partial\Phi_m}{\partial r} + \frac{\partial^2\Phi_m}{\partial z^2} - \frac{m^2\Phi_m}{r^2} = 0 \qquad (2.42)$$

If we now expand the functions $\Phi_m(r,z)$ as a power series in r, of the general form

$$\Phi_m(r,z) = \sum_{n=0,1,2...}^{\infty} c_{mn}(z)r^n \qquad (2.43)$$

Then the relationship between the functions $c_{mn}(z)$ can readily be deduced by invoking the fact that $\Phi_m(r,z)$ must be a solution to Laplace's equation. So the expressions for $\Phi_1(r,z)$ and $\Phi_3(r,z)$ can be found to have the general form

$$
\left.\begin{array}{l}
\Phi_1(r,z) = d_1(z)r - \dfrac{1}{8}d_1''(z)r^3 + ... \\[2mm]
\Phi_3(r,z) = d_3(z)r^3 - \dfrac{1}{16}d_3''(z)r^5 + ...
\end{array}\right\}
\tag{2.44}
$$

The general expression for $\Phi_m(r,z)$ is then given by

$$
\Phi(r,z) = \left[d_1(z)r - \tfrac{1}{8}d_1''(z)r^3\right]\cos\theta + d_3(z)r^3 \cos 3\theta + \cdots
\tag{2.45}
$$

This equation shows that the functions $d_1(z)$ and $d_3(z)$ completely define the deflection potential near the optic axis, up to third-order powers of the off-axis distance r. Therefore, if the function $d_1(z)$ and $d_3(z)$ are known for any deflector, then its third-order aberrations properties can be calculated.

It is necessary to derive the harmonic relationship between the vector function **F** and winding distribution of saddle and toroidal coils, as well as the harmonic boundary condition for electrodes. The following formulation of the vector function **F** is based upon work reported by Munro and Chu [4].

4.3   Source term for Toroidal Coils

Figure 2.10a shows a simple toroidal yoke in the presence of a rotationally symmetric magnetic core. The coil windings have a mean semi-angle φ, and each coil has an excitation of NI ampere-turns, uniformly distributed over a small angle of 2δ about the mean semi-angle φ. The current distribution in the yoke can be expressed as a Fourier series of harmonic components. In practice, a current loading function $f(\theta)$, defined as the angular ampere-turn distribution, as illustrated in Figure 2.10b, is used and expressed in the form

$$
f(\theta) = \sum_{m=1,3,5,...}^{\infty} f_m \sin m\theta \qquad \text{where} \qquad f_m = \frac{4NI}{\pi}\sin m\varphi
\tag{2.46}
$$

The angular thickness of the coil (2δ) is assumed to be small. The next step is to relate the harmonic expression of vector function **F** to the current loading function. Toroidal coils, as shown in Figure 2.10, are normally wound in the radial plane. The exciting current **J** has only r and z components, but no θ component. It therefore follows that the function **F** has only a θ-component. It can be shown that

$$
F_\theta(r,z,\theta) = \frac{1}{r}g(r,z)f(\theta)
\tag{2.47}
$$

where the function g(r,z) is unity inside the coil windings and zero elsewhere.

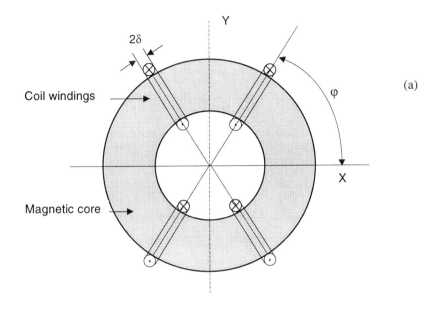

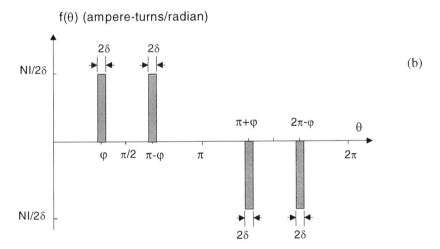

Figure 2.10   Driving current for toroidal coils

(a) End view of a toroidal yoke in the presence of rotationally symmetric magnetic core (only the H_x-coils are shown)

(b) Current loading function for the toroidal yokes

4.4   Source term for Saddle Coils

The deflection field functions for a saddle yoke in the presence of a rotationally symmetric magnetic material can be computed in a similar way to that described in the previous section. Figure 2.11a shows the end view of a saddle yoke, with mean semi-angle φ and each coil having an excitation of NI ampere-turns uniformly distributed over a small distance ΔR perpendicular to the flow of current.

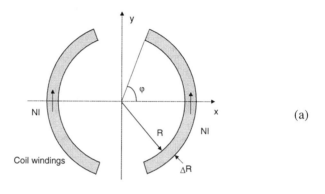

(a)

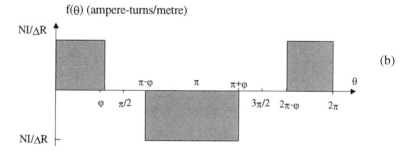

(b)

Figure 2.11 Driving current for Saddle coils
(a)   End view of a saddle yoke
(b)   Current loading function

The current loading function f(θ), in ampere-turns/metre, is shown in Figure 2.11b and can be expressed as a Fourier series of the form

$$f(\theta) = \sum_{m=1,3,5,...}^{\infty} f_m \cos m\theta \qquad \text{where} \qquad f_m = \frac{4NI}{\pi m \Delta R} \sin m\varphi \qquad (2.48)$$

The vector function **F** can be related to the current source by

$$F_r(r,z,\theta) = g(r,z)f(\theta)\cos\alpha$$
$$F_z(r,z,\theta) = -g(r,z)f(\theta)\sin\alpha$$

(2.49)

where α is the deflector taper angle relative to the z direction and $g(r,z)$ is equal to unity inside the coil windings and zero elsewhere.

4.5   Boundary conditions on Multipole Electrodes

To obtain the appropriate harmonic boundary condition on electrostatic deflection electrodes, assume $U(\theta)$ to be the potential distribution on the rotational surface containing the electrodes. Then $U(\theta)$ can be expressed as a Fourier series of the general form

$$U(\theta) = \sum_{m=1,3,5..} \Phi_{Bm} \cos m\theta \quad \text{where} \quad \Phi_{Bm} = \frac{4}{\pi} \int_0^{\pi/2} U(\theta)\cos m\theta d\theta$$

(2.50)

where Φ_{Bm} is the harmonic boundary condition on the electrode. In order to evaluate this integral, we make an approximation that on the electrode rotational surface, the potential varies linearly as a function of θ in the gap between adjacent electrodes. With this approximation, the function $U(\theta)$ has the piecewise linear form as shown in Figure 2.12.

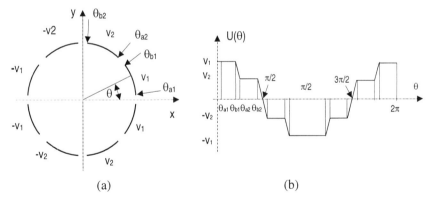

(a)   (b)

Figure 2.12 Electrodes on a cylindrical surface
(a) End-view
(b) Assumed potential distribution $U(\theta)$ on the cylindrical surface

The integral in equation (2.50) can be evaluated analytically to give

$$\Phi_{Bm} = \frac{4}{\pi m^2}\left\{\left[\sum_{i=1}^{n-1}(V_{i+1}-V_i)\left(\frac{\cos m\theta_{a_{i+1}}-\cos m\theta_{b_i}}{\theta_{a_{i+1}}-\theta_{b_i}}\right)\right]+V_n\left(\frac{\cos m\theta_{b_n}}{\pi/2-\theta_{b_n}}\right)\right\} \quad (2.51)$$

where n denotes the number of electrodes in the first quadrant, V_i denotes the potential of the ith electrode and θ_{ai} and θ_{bi} defines the angular extent of the ith electrode (see Figure 2.12).

REFERENCES

[1]   P. W. Hawkes and E. Kasper, "Principles of Electron Optics", Volume 1, Basic Geometrical Optics, Academic Press, 1989, London

[2]   Janming Jin, "The Finite Element Method in Electromagnetics", John Wiley and Sons, Inc., New York, 1993

[3]   P. Grivet, "Electron Optics", Revised by A. Septier, translated by P. W. Hawkes, Permagon Press, 1965, London

[4]   E. Munro, and H.C Chu, "Numerical analysis of electron beam lithography systems. Part I: computation of fields in magnetic deflectors", Optik, 60, pp.371-390. 1982 and their paper, "Numerical Analysis of Electron beam systems. Part II: Computation of fields in electrostatic deflectors", Optik, 61,1-16, 1982

Chapter 3
THE FINITE DIFFERENCE METHOD

The finite difference method was traditionally used in electron optics for solving field distributions. Even for magnetic field calculations, where the finite element method has largely replaced it, there are instances where the finite difference method is still advocated [1]. Finite elements are closely related to finite differences, indeed, there are instances where the two are equivalent. The same type of error analysis and matrix solution methods apply to them both. For these reasons, a brief description of the finite difference method will be given here.

A simple finite difference mesh for an electron lens is given in Figure 3.1. V_1 and V_2 at the pole pieces can either represent electric potentials or magnetic scalar potentials.

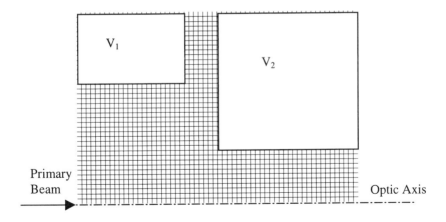

Figure 3.1 A finite difference mesh for an electron lens

The layout depicted in Figure 3.1 is a relatively simple one, the mesh spacing is regular and the pole-pieces lie on mesh lines. Laplace's equation is solved for potential values at the mesh nodes.

1.   LOCAL FINITE 5PT DIFFERENCE EQUATIONS

Consider a two dimensional domain defined in Cartesian coordinates (x,y). The finite difference method starts by dividing up the domain into a mesh with regular spacing h, where a typical mesh node 0 is surrounded by four other nodes: 1, 2, 3, 4, as shown in Figure 3.2

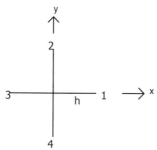

Figure 3.2 A 5pt Finite Difference Star

A Taylor's expansion about the centre node can obviously be made in the x and y directions. For points 1 and 3

$$V_1 = V_0 + h(\frac{\partial V}{\partial x})_0 + \frac{h^2}{2!}(\frac{\partial^2 V}{\partial x^2})_0 + \frac{h^3}{3!}(\frac{\partial^3 V}{\partial x^3})_0 + \ldots$$

$$V_3 = V_0 - h(\frac{\partial V}{\partial x})_0 + \frac{h^2}{2!}(\frac{\partial^2 V}{\partial x^2})_0 - \frac{h^3}{3!}(\frac{\partial^3 V}{\partial x^3})_0 + \ldots$$

(3.1)

Adding these two equations together and neglecting terms above second order

$$V_1 + V_3 - 2V_0 = h^2(\frac{\partial^2 V}{\partial x^2})_0$$

(3.2)

and in the y direction

$$V_2 + V_4 - 2V_0 = h^2(\frac{\partial^2 V}{\partial y^2})_0$$

(3.3)

substituting into Laplace's equation gives

$$V_1 + V_2 + V_3 + V_4 - 4V_0 = 0$$

(3.4)

The voltage on the centre node is obviously the average of the surrounding nodes.

For Poisson's equation, where Q is the charge at the centre node

$$V_1 + V_2 + V_3 + V_4 - 4V_0 = h^2 Q$$

(3.5)

This equation applies to every node in the domain. Note that the finite difference equation corresponding to Laplace's equation is independent of h, the mesh spacing distance.

The general strategy is to start with a partial differential equation, and substitute the derivatives with numerical approximations that are expressed in terms of local potential values. The substitution for second-order derivatives is given by

$$\left(\frac{\partial^2 V}{\partial x^2}\right)_0 = \left(\frac{V_3 + V_1 - 2V_0}{h^2}\right) \qquad \left(\frac{\partial^2 V}{\partial y^2}\right)_0 = \left(\frac{V_2 + V_4 - 2V_0}{h^2}\right) \qquad (3.6)$$

For Laplace's equation in other coordinate systems, consider the more general equation

$$a\frac{\partial^2 V}{\partial x^2} + b\frac{\partial^2 V}{\partial y^2} + c\frac{\partial V}{\partial x} + d\frac{\partial V}{\partial y} = 0 \qquad (3.7)$$

where a, b, c, and d are constants. The two first-order derivatives are approximated by the following central difference expressions

$$\left(\frac{\partial V}{\partial x}\right)_0 = \left(\frac{V_1 - V_3}{2h}\right) \qquad \left(\frac{\partial V}{\partial y}\right)_0 = \left(\frac{V_2 - V_4}{2h}\right) \qquad (3.8)$$

The finite difference nodal equation for this situation is given by

$$V_1\left(a + \frac{hc}{2}\right) + V_2\left(b + \frac{hd}{2}\right) + V_3\left(a - \frac{hc}{2}\right) + V_4\left(b - \frac{hd}{2}\right) - 2V_0(a + b) = 0 \qquad (3.9)$$

Note that for a=b=1 and c=d=0, this equation is identical to the former equation. Note also that a, b, c and d can be functions of x and y, a(x,y), b(x,y), c(x,y) and d(x,y). When this equation is applied, the coordinates of the node 0 is then used to determine a, b, c and d.

The finite difference representation of Laplace's equation in the (x,y) plane can be represented in the following pictorial form

$$\nabla^2 u_{ij} = \frac{1}{h^2}\begin{Bmatrix} & 1 & \\ 1 & -4 & 1 \\ & 1 & \end{Bmatrix}u \qquad (3.10)$$

2.   THE MATRIX EQUATION

Consider the problem where the values of a function u(x,y) are explicitly defined at the sides of a box, as shown in Figure 3.3.

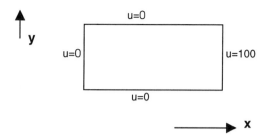

Figure 3.3 A simple electrostatic box example

Using a very small number of mesh nodes to illustrate the method of solving for u(x,y) inside the box, we use the mesh layout shown in Figure 3.4.

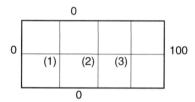

Figure 3.4 A three-node finite difference mesh

The free nodes are located at points (1), (2) and (3). At all other points, the value of u(x,y) are fixed. Assume these nodes are equally spaced. Let the value of u(x,y) at these nodes be denoted by u_1, u_2, and u_3 respectively, so that three difference equations are generated

node 1: $0 + 0 + u_2 + 0 - 4u_1 = 0$
node 2: $u_1 + 0 + u_3 + 0 - 4u_2 = 0$
node 3: $u_2 + 0 + 100 + 0 - 4u_3 = 0$ (3.11)

This can obviously be put in matrix form

$$\begin{bmatrix} -4 & 1 & \\ 1 & -4 & 1 \\ & 1 & -4 \end{bmatrix} \begin{bmatrix} u_1 \\ u_2 \\ u_3 \end{bmatrix} = \begin{bmatrix} 0 \\ 0 \\ -100 \end{bmatrix}$$ (3.12)

The solution of this can easily be done algebraically, and $u_1 = 1.786$, $u_2 = 7.143$, $u_3 = 26.786$. For the box example, the exact solution can be found through the Fourier series method [2], and the exact values for [u] are (1.0943, 5.4885, 26.0944). The percentage errors on nodes 1, 2 and 3 are 63.20%, 30.14%, and 2.65% respectively.

For n unknown nodes, the final matrix equation may be stated in the following form

$$[A][u] = [B] \tag{3.13}$$

where [A] is of rank n by n, [B] is a column vector of length n, and [u] is a column vector of the unknown potentials. The number of nodes typically range from 1000 to 100000, and numerical techniques are used to solve the above matrix equation. The matrix [A] is "sparse", that is, it contains relatively few non-zero terms. For 5pt finite difference in two-dimensions, [A] has the following form

$$
\begin{bmatrix}
-4 & 1 & & & 1 & & & & \\
1 & -4 & 1 & & & 1 & & & \\
& 1 & -4 & 1 & & & 1 & & \\
& & 1 & -4 & 1 & & & 1 & \\
1 & & & 1 & -4 & 1 & & & 1 \\
& 1 & & & 1 & -4 & 1 & & & 1 \\
\cdot & \cdot & \cdot & \cdot & \cdot & \cdot & \cdot & \cdot & \cdot \\
\cdot & \cdot & \cdot & \cdot & \cdot & \cdot & \cdot & \cdot & \cdot \\
& & & & 1 & & & 1 & -4 & 1 \\
& & & & & 1 & & & 1 & -4
\end{bmatrix}
$$

The above linear system of equations can either be solved by direct methods such as Gaussian elimination, or by iterative methods such as successive overrelaxation and conjugate gradient algorithms, or any of their numerous variations [3]. An efficient and fast technique commonly used to solve the above system of equations is the Incomplete Factorization of the Conjugate Gradient Method (ICCG) [4]. The Finite element method also generates a sparse set of linear equations, very similar to the finite difference one. The same matrix solution methods are used for both techniques.

3.   TRUNCATION ERRORS

The truncation error is the most important type of error in the finite difference method. Rounding errors are usually small due to the high number of digits that are used to represent numbers in modern computers.

Truncation errors can be estimated by examining the Taylor's series for the finite difference star. Consider the 5 pointed star shown in Figure 3.2

$$V_1 + V_3 = 2V_0 + h^2 \left(\frac{\partial^2 V}{\partial x^2} \right)_0 + \frac{h^4}{12} \left(\frac{\partial^4 V}{\partial x^4} \right)_0 + \cdots$$

$$V_2 + V_4 = 2V_0 + h^2 \left(\frac{\partial^2 V}{\partial y^2} \right)_0 + \frac{h^4}{12} \left(\frac{\partial^4 V}{\partial y^4} \right)_0 + \cdots \tag{3.14}$$

where V(x,y) is the electric potential. Laplace's equation in finite difference form is

$$\nabla^2 V = \frac{1}{h^2} (V_1 + V_3 + V_2 + V_4 - 4V)_0 - \frac{h^2}{12} \left(\frac{\partial^4 V}{\partial x^4} + \frac{\partial^4 V}{\partial y^4} \right)_0 + \cdots \tag{3.15}$$

The truncation error $T(x,y,h)$ is given by

$$T(x, y, h) \geq \frac{h^2}{12} \left(\frac{\partial^4 V}{\partial x^4} + \frac{\partial^4 V}{\partial y^4} \right)_0 \qquad (3.16)$$

Note the dependence of $T(x,y,h)$ on the mesh size h: if h is halved, then the truncation error is reduced by a factor of 4. Also note the dependence of the truncation error on the 4th derivative of the potential. This means that the truncation error is low for smoothly varying fields, while discontinuities in the field distribution will sharply increase it.

Where the dependence of the truncation error on mesh spacing is known, it can be significantly reduced by using an extrapolation procedure. Consider two potential solutions: V_h for a mesh size of h, and $V_{h/2}$ for a mesh size of h/2. Let V be the true solution, then obviously

$$V - V_h = gh^2$$
$$V - V_{h/2} = \frac{1}{4} gh^2 \qquad (3.17)$$

where g is a constant that can be eliminated to give a more accurate solution

$$V = \frac{4}{3} V_{h/2} - \frac{1}{3} V_h \qquad (3.18)$$

This leads to a significant reduction of the error. It can readily be shown that the truncation error is now proportional to the forth power of h. The method is known as Richardson's extrapolation technique and it is also used on finite element solutions.

4.   ASYMMETRIC STARS

In the finite difference method, because the potential is only defined at the mesh nodes, the situation where electrodes or boundaries cross the mesh at points other than mesh nodes is a common occurrence and needs special treatment. In Figure 3.5, a curved boundary intersects the mesh at nodes 1 and 2, defined by the distances ph and qh respectively from the centre node 0.

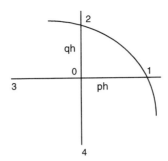

Figure 3.5 An asymmetrical finite difference star

To find the appropriate nodal equation, a Taylor's expansion can be used as before, only this time, the arms of the finite difference star are reduced to ph in the x direction, and qh in the y-direction. By suitably weighting the potentials, an expression for the second spatial derivative of the potential in either direction can be found. Consider the potentials in the x-direction

$$V_1 = V_0 + ph\left(\frac{\partial V}{\partial x}\right)_0 + \frac{p^2 h^2}{2!}\left(\frac{\partial^2 V}{\partial x^2}\right)_0 + \ldots\ldots$$

$$V_3 = V_0 - h\left(\frac{\partial V}{\partial x}\right)_0 + \frac{h^2}{2!}\left(\frac{\partial^2 V}{\partial x^2}\right)_0 + \ldots\ldots$$

$$V_1 + pV_3 = (1+p)V_0 + p(p+1)\frac{h^2}{2!}\left(\frac{\partial^2 V}{\partial x^2}\right)_0 + \ldots\ldots$$

(3.19)

by forming a similar equation in the y direction, the second order derivatives are

$$\left(\frac{\partial^2 V}{\partial x^2}\right)_0 = \frac{2V_1}{h^2 p(p+1)} + \frac{2V_3}{h^2(p+1)} - \frac{2V_0}{h^2 p}$$

$$\left(\frac{\partial^2 V}{\partial y^2}\right)_0 = \frac{2V_2}{h^2 q(q+1)} + \frac{2V_4}{h^2(q+1)} - \frac{2V_0}{h^2 q}$$

(3.20)

so that the nodal equation for Laplace's equation becomes

$$2\left[\frac{V_1}{p(1+p)} + \frac{V_2}{q(1+q)} + \frac{V_3}{(1+p)} + \frac{V_4}{(1+q)} - \left(\frac{1}{p}+\frac{1}{q}\right)V_0\right] = 0$$

(3.21)

Note that this nodal equation is independent of the mesh spacing h. At every node which has an asymmetric star, the above type of nodal equation must be used.

The truncation error can be derived by retaining higher-order terms in the Taylor's series, and is given by

$$T(x,y) \geq \frac{h}{3}\left((p-1)\left(\frac{\partial^3 V}{\partial x^3}\right)_0 + (q-1)\left(\frac{\partial^3 V}{\partial y^3}\right)_0\right) + \frac{h^2}{12}\left(\frac{(p^3+1)}{(p+1)}\left(\frac{\partial^4 V}{\partial x^4}\right) + \frac{(q^3+1)}{(q+1)}\left(\frac{\partial^4 V}{\partial y^4}\right)\right)$$

$$(3.22)$$

This is an important expression not only for finite difference, but also for the finite element method. It turns out that in many instances, the nodal equation and truncation error for the two methods are identical. Here, $T(x,y)$ is not only proportional to h^2, but it also varies with h. This means that where the boundary intersects off-nodal mesh points, the result will not be as accurate as for a mesh which has symmetric stars.

Figure 3.6 depicts the more general situation where a finite difference star has unequal arm lengths on all sides.

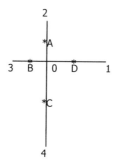

Figure 3.6 General asymmetric star

Let h_1 be the length of the right hand side arm, h_3 be left hand side arm length, h_2 be the top arm length, and h_4 be the bottom arm length. Let A, B, C and D be points that bisect each arm. A first-order approximation for the spatial derivatives at the midpoint of each arm in the x direction are given by

$$\left(\frac{\partial u}{\partial x}\right)_D = \frac{V_1 - V_0}{h_1} \qquad \left(\frac{\partial u}{\partial x}\right)_B = \frac{V_0 - V_3}{h_3} \qquad (3.23)$$

Since the distance between points B and D is $(0.5h_1 + 0.5h_3)$ then

$$\frac{\partial^2 V}{\partial x^2} = \frac{\partial}{\partial x}\left(\frac{\partial V}{\partial x}\right) = \frac{2}{(h_1+h_3)}\left[\left(\frac{\partial V}{\partial x}\right)_D - \left(\frac{\partial V}{\partial x}\right)_B\right] \qquad (3.24)$$

Using the same reasoning, a similar expression can be obtained in the y-direction

$$\frac{\partial^2 V}{\partial y^2} = \frac{\partial}{\partial y}\left(\frac{\partial V}{\partial y}\right) = \frac{2}{(h_2+h_4)}\left[\left(\frac{\partial V}{\partial x}\right)_A - \left(\frac{\partial V}{\partial x}\right)_C\right] \qquad (3.25)$$

It is straightforward to show that when the above double derivatives are substituted into Laplace's equation, the following 5pt difference star is obtained

$$2\begin{Bmatrix} 0 & \dfrac{1}{h_2(h_2+h_4)} & 0 \\[2mm] \dfrac{1}{h_3(h_1+h_3)} & -\left(\dfrac{1}{h_1 h_3}+\dfrac{1}{h_2 h_4}\right) & \dfrac{1}{h_1(h_1+h_3)} \\[2mm] 0 & \dfrac{1}{h_4(h_2+h_4)} & 0 \end{Bmatrix}V=0 \qquad (3.26)$$

For $h_2=h_4=h$ and $h_1=h_3=h/2$, that is, where the horizontal arms are half the size of the vertical arms, the picture star simplifies to

$$\dfrac{2}{h^2}\begin{Bmatrix} & 0.5 & \\ 2 & -5 & 2 \\ & 0.5 & \end{Bmatrix}V=0 \qquad (3.27)$$

This shows that the influence of nodes 2 and 4 is halved, while that of nodes 1 and 3 is doubled. Note that the sum of the off-diagonal entries is equal to the magnitude of the diagonal entry. The total sum of all entries in the picture star is zero. This property is true for all finite difference and finite element nodal equations that relate to field problems.

5.   MATERIAL INTERFACES

Consider two regions of different dielectric permittivity, regions A and B as shown in Figure 3.7

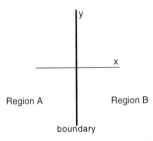

Figure 3.7 Regions of different dielectric permittivity

The boundary conditions at the interface require continuity of the tangential electric field

$$\frac{\partial V_a}{\partial y} = \frac{\partial V_b}{\partial y}$$
(3.28)

and normal electric flux

$$\varepsilon_a \frac{\partial V_a}{\partial x} = \varepsilon_b \frac{\partial V_b}{\partial x}$$
(3.29)

Now let there be a finite difference star at the interface, where nodes 2, 0 and 4 lie on the interface, as shown in Figure 3.8.

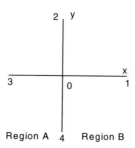

Figure 3.8 Finite Difference at a boundary interface

The finite difference form of Laplace's equation for region A is given by

$$V_{1a} + V_{2a} + V_{3a} + V_{4a} - 4V_{0a} = 0$$
(3.30)

Similarly for region B

$$V_{1b} + V_{2b} + V_{3b} + V_{4b} - 4V_{0b} = 0$$
(3.31)

Points V_{3b} and V_{1a} are fictitious points that can be eliminated by applying the boundary conditions. Obviously for the potentials that lie on the boundary

$$V_{2a} = V_{2b} = V_2, \quad V_{4a} = V_{4b} = V_4, \quad V_{0a} = V_{0b} = V_0$$
(3.32)

while continuity of the normal component of the flux density D_x gives

$$\varepsilon_a (V_{1a} - V_{3a}) = \varepsilon_b (V_{1b} - V_{3b})$$
(3.33)

We can eliminate the fictitious points V_{1a} and V_{3b} by using the finite difference form of Laplace's equation for each region separately, so that

$$V_{1a} = -V_2 - V_{3a} - V_4 + 4V_0, \quad V_{3b} = -V_2 - V_4 - V_{1b} + 4V_0$$
(3.34)

Substituting back into the normal boundary condition this gives the nodal equation for the interface

$$2V_{1b} + (1+S)V_2 + 2SV_{3a} + (1+S)V_4 - 4(1+S)V_0 = 0 \quad \text{where } S = \frac{\varepsilon_a}{\varepsilon_b} \quad (3.35)$$

The pictorial star is given by

$$\begin{Bmatrix} 0 & 1+S & 0 \\ 2S & -4(1+S) & 2 \\ 0 & 1+S & 0 \end{Bmatrix} V \qquad (3.36)$$

6.   THE NINE POINTED STAR IN RECTILINEAR COORDINATES

An effective way of reducing the truncation error is to use a nine pointed difference star, as shown in Figure 3.9

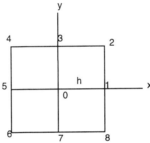

Figure 3.9 A Nine pointed Star

The coefficients which multiply the potentials, $V_0....V_8$ in the nodal equation can be chosen to cancel higher-order terms in the two-dimensional Taylor's series expansion of the potential. In this case, the truncation error is limited by $O(h^6)$ terms. The nine point finite difference pictorial star for Laplace's equation is given by

$$\nabla^2 V = \frac{1}{6h^2} \begin{Bmatrix} 1 & 4 & 1 \\ 4 & -20 & 4 \\ 1 & 4 & 1 \end{Bmatrix} V = 0 \qquad (3.37)$$

This star can be broken down into the weighted average of two five pointed stars: one for a 45 degree tilted star with the nodes 2, 4, 6, 8 at its vertices, and the other for the normal five pointed star with nodes 1, 3, 5, 7 at its vertices, that is,

$$\nabla^2 V = \frac{2}{3h^2} \begin{Bmatrix} \begin{bmatrix} 0 & 1 & 0 \\ 1 & -4 & 1 \\ 0 & 1 & 0 \end{bmatrix} V \end{Bmatrix} + \frac{1}{6h^2} \begin{Bmatrix} \begin{bmatrix} 1 & 0 & 1 \\ 0 & -4 & 0 \\ 1 & 0 & 1 \end{bmatrix} V \end{Bmatrix} \qquad (3.38)$$

Consider the box problem shown in Figures 3.3 and 3.4 for the nine pointed star. Note that the corner potentials of the box are now required. For the meeting of the two plates on the right hand side, take the average potential to be 50V.

The nine pointed star gives the following matrix equation

$$\begin{pmatrix} -5 & 1 & 0 \\ 1 & -5 & 1 \\ 0 & 1 & -5 \end{pmatrix} \begin{pmatrix} V_1 \\ V_2 \\ V_3 \end{pmatrix} = \begin{pmatrix} 0 \\ 0 \\ -125 \end{pmatrix} \qquad (3.39)$$

The solution to this equation is, $V_1 = 1.0895$, $V_2 = 5.4347$, and $V_3 = 26.0869$. A comparison of the percentage accuracy of these values with 5pt finite difference is given in Table 3.1

	V_1	V_2	V_3
Analytical values	1.09433	5.4885	26.0944
Accuracy of 5pt star	63.2%	30.14%	2.65%
Accuracy of 9pt star	0.44%	0.98%	0.029%

Table 3.1 Comparison of the accuracy of the 5 pt and 9pt stars for the 3 node Box problem

Clearly the accuracy of the 9 pt finite difference method is around one to two orders of magnitude more accurate than the 5 pt star. In general, the advantage of the 9pt finite difference star depends on the problem under study. In this case, the factor of improvement is somewhat larger than usual due to the existence of singularities at the right-hand side corners of the box. Nevertheless, these results clearly illustrate the advantage of using the 9 pt star. In charged particle optics, Kasper has formulated 9 pt finite difference for a variety of different coordinate systems [5].

7.   AXISYMMETRIC CYLINDRICAL COORDINATES

Consider the case of axisymmetry, where the potential is constant in the azimuthal direction, so that $V(r,\phi,z) = V(r,z)$. Laplace's equation is simplified to

$$\nabla^2 V = \frac{\partial^2 V}{\partial r^2} + \frac{1}{r}\frac{\partial V}{\partial r} + \frac{\partial^2 V}{\partial z^2} = 0 \tag{3.40}$$

Solving this equation is a two dimensional field problem in the (r,z) plane. The 5pt finite difference star can be used, as shown in Figure 3.10a and 3.10b.

(a)   (b)

Figure 3.10 Finite difference star in (r,z)

The finite difference approximations are given by

$$\frac{\partial V}{\partial r} = \frac{(V_2 - V_4)}{2h} \quad \frac{\partial^2 V}{\partial r^2} = \frac{(V_2 + V_4 - 2V_0)}{h^2} \quad \frac{\partial^2 V}{\partial z^2} = \frac{(V_1 + V_3 - 2V_0)}{h^2} \tag{3.41}$$

substituting these approximations into Laplace's equation gives

$$\frac{V_2 + V_4 - V_0}{h^2} + \frac{V_2 - V_4}{2hr_0} + \frac{V_1 + V_3 - V_0}{h^2} = 0 \tag{3.42}$$

where r_0 is the radius at node 0. This expression can be simplified into the following nodal equation

$$V_1 + V_3 + \left(1 + \frac{h}{2r_0}\right)V_2 + \left(1 - \frac{h}{2r_0}\right)V_4 - 4V_0 = 0 \tag{3.43}$$

There are obvious difficulties on the axis when $r_0=0$. Since the potential around the axis must be an even function in r, then

$$V(r) = a_0 + a_2 r^2 + a_4 r^4 + \ldots\ldots\ldots \tag{3.44}$$

then

$$\frac{1}{r}\frac{\partial V}{\partial r} = \frac{\partial^2 V}{\partial^2 r} \quad r \to 0 \tag{3.45}$$

This in turn leads to the following pictorial star for on-axis nodes.

$$(1/h^2)\left\{\begin{array}{ccc} & 4 & \\ 1 & -6 & 1 \end{array}\right\}V \tag{3.46}$$

Consider the closed tube conductor layout shown in Figure 3.11. The potential distribution inside the tube can be solved analytically by using a Fourier-Bessel series [6].

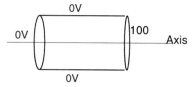

Figure 3.11 Example closed tube problem

Take six evenly spaced free nodes, as shown in Figure 3.12

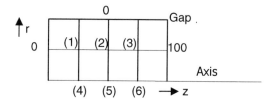

Figure 3.12 Six free node finite difference mesh for closed tube problem

Applying the nodal equations given by (3.43) and (3.46), a 5pt difference finite difference scheme for the closed tube problem yields the following matrix equation

$$
\begin{pmatrix}
-4 & 1 & 0 & \frac{1}{2} & 0 & 0 \\
1 & -4 & 1 & 0 & \frac{1}{2} & 0 \\
0 & 1 & -4 & 0 & 0 & \frac{1}{2} \\
4 & 0 & 0 & -6 & 1 & 0 \\
0 & 4 & 0 & 1 & -6 & 1 \\
0 & 0 & 4 & 0 & 1 & -6
\end{pmatrix}
\begin{pmatrix}
V_1 \\ V_2 \\ V_3 \\ V_4 \\ V_5 \\ V_6
\end{pmatrix}
=
\begin{pmatrix}
0 \\ 0 \\ -100 \\ 0 \\ 0 \\ -100
\end{pmatrix}
\tag{3.47}
$$

The values of the potentials on the optic axis after solving this equation are, $V_4=4.6905$, $V_5=14.8936$, $V_6=41.0541$ which when compared to the analytical solution give respective errors of 19.44%, 6.85% and 2.98%.

By considering higher-order terms in Taylor's series, it is straightforward to show that for constant mesh spacing of h, the truncation error in axisymmetric cylindrical coordinates is given by

$$T(r,z) \geq \frac{h^2}{12}\left(\frac{\partial^4 V}{\partial z^4} + \frac{\partial^4 V}{\partial r^4} + \frac{2}{r}\frac{\partial^3 V}{\partial r^3}\right) \qquad (3.48)$$

This expression is important. It indicates that the truncation error varies with h^2, so Richardson's extrapolation method is effective. On the other hand, the presence of the 1/r term means that the truncation error will be high for nodes close to the optic axis. For nodes on the optics axis, the truncation is error is given by

$$T(r,z) \geq \frac{h^2}{12}\left(\frac{\partial^4 V}{\partial z^4} + 2\frac{\partial^4 V}{\partial r^4}\right) \qquad (3.49)$$

which is obviously very similar to the one derived for (x,y) rectilinear coordinates.

The nine point star nodal equation for axi-symmetric coordinates has been derived by Kasper [7]. Consider the 6-node closed tube problem depicted in Figure 3.12. The 9 point star approximation to this mesh layout provide the following potentials on the optic axis, $V_4=3.9014$, $V_5=13.8925$, $V_6=42.3055$. Table 3.2 compares the accuracy of these potentials with 5pt star finite difference.

	V_4	V_5	V_6
Analytical values	3.92686	13.93372	42.31812
Accuracy of 5pt star	19.44%	6.85%	2.98%
Accuracy of 9pt star	0.64%	0.29%	0.030%

Table 3.2 Comparison of the accuracy of the 5 pt and 9pt stars for the 6 node closed tube problem

The above results show that just as in the rectilinear coordinate case, the use of 9 pt finite difference improves the accuracy by over an order of magnitude.

REFERENCES

[1]   J. Rouse and E. Munro, "Three-dimensional computer modelling of electrostatic and magnetic electron optical components", J. Vacuum Science and Technology, B7(6), 1989, p1891-1897

[2]   E. Kreyszig, "Advanced Engineering Mathematics", 7th edition, John Wiley and Sons, 1993, New York, p651-3

[3]   G. F. Carey and J. T. Oden, "Finite Elements: Computational Aspects III", Prentice-Hall Inc., New Jersey USA, 1984

[4]   A widely used method is given by A. Meijerink and H. A. van der Vorst in their paper entitled, "An iterative solution method for linear systems of which the coefficient matrix is a symmetric M-matrix", Mathematics of Computation, Jan. 1977, p148-162.   This method is suitable for situations where the non-zeros have a fixed diagonal pattern. This typically occurs for boundary conditions at the outer boundaries of the whole domain. But where there are internal boundaries, the method presented by N. Munksgaard in his paper, "Solving sparse symmetric sets of linear equations by preconditioned conjugate gradients", ACM Transactions on Mathematical Software, Vol. 6, No. 2, June 1980, p206-219 is more suitable. It can deal with unstructured non-zero patterns and is in general, much more flexible.

[5]   See for instance, section 11.4 (p165-170), in "Principles of Electron Optics", by P. W. Hawkes and E. Kasper

[6]   S. Ramo, J. R. Whinnnery, and T. Van Duzer, "Fields and waves in communication electronics", John Wiley and Sons, second edition, Newy York, 1984, p373-5

[7]   E. Kasper, "Field calaculation and the determination of electron trajectories", chapter 2, p72-3, in the book, "Magnetic Electron Lenses", edited by P. W. Hawkes, Springer-Verlag, New York, 1982

Chapter 4
FINITE ELEMENT CONCEPTS

There are a number of difficulties with the finite difference method which are overcome by the finite element method. Finite difference methods need to use unequal stars close to curved boundaries. This results in an asymmetric final matrix that slows down convergence to a stable solution. It also involves more complicated programming, since the nodal equation for each node close to the boundary is different and unique to the boundary shape.

There is also the issue of curved boundaries with derivative boundary conditions. In this case, not only are asymmetrical stars required, but additional equations at the boundary must be enforced which involve first-order derivatives of the potential. This involves a complicated procedure of constructing fictitious points and then eliminating them. The whole procedure, although possible, is complicated and cumbersome.

Often high field strengths are localised to specific regions of the domain. Obviously the most efficient approach is to refine the mesh locally. Although grading of the finite difference mesh is possible, it can only be done separately for each direction. In practice, mesh refinement in finite difference is restricted to the situation where regions of high field strengths are clearly demarcated from one another.

In cases where different material regions exist, the finite difference method needs to alter the nodal equation at every node that lies on the interface between different materials, so that the correct boundary conditions are satisfied. This requires enforcing a constraint on the normal derivative of the field at the interface, and is usually done by introducing fictitious points within each region, and then eliminating them. If the boundary interface is curved, this is a complicated procedure. The finite element overcomes many of these difficulties. It is of course, not impossible to use the finite difference method for the situations just described, but it is cumbersome when compared to the finite element approach.

1.   FINITE ELEMENTS IN ONE DIMENSION

1.1   The weighted residual approach

Consider an electrostatic field distribution which has no free charge over the domain R. The divergence of the electric flux vector $\mathbf{D}(x,y)$ is zero at every point within the domain

$$\nabla \cdot \mathbf{D} = 0 \qquad\qquad (4.1)$$

The finite difference method attempts to satisfy this equation only at mesh nodes on a regular mesh. The total domain R_T is in effect, approximated by a network of selected points, as shown in Figure 4.1.

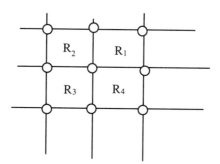

Figure 4.1 Network Mesh model

The finite difference method is formulated in terms of a potential $u_{i,j}$ which is only defined at the points $(x_{i,j}, y_{i,j})$ where the indices i and j denote mesh node numbers. At each node, there is a certain precision to which the equation is solved

$$\nabla^2 u = e \qquad\qquad (4.2)$$

where e is the residual error at each mesh node. The finite difference method proceeds by explicitly enforcing e to be zero at each (i,j). But what about the space in between? Is there any way we can also attempt to satisfy the equation over the regions, R_1, R_2 R_3, and R_4, in the space in between the nodes?

Now consider an alternative approach. Let us define an integrated form of the residual e_R in the space around a mesh node, over regions R_1, R_2 R_3, R_4 which we denote by R, so that

$$e_R = \iint_R v(\nabla \cdot \mathbf{D}) dxdy = 0 \qquad\qquad (4.3)$$

Where $v(x,y)$ is a "hat" function which is equal to unity at the node (i,j), and is zero at the surrounding nodes. It may be thought of as a normalised weighting factor.

Note how this residual involves the regions around the node. We can enforce this residual to be zero, and the finite element does exactly this. The weighting function describes how to localise the residual error. Note that in the finite difference method, $v(x,y)$ is effectively a delta function at the mesh node coordinates, confining the residual to mesh nodes. Viewed in this way, the finite difference method is a subset of the weighted residual method.

What is the advantage of defining a weighted average of the residual over the region surrounding a node? To understand this, let us first consider the simple one dimensional form of Gauss's law in electrostatics

$$e_R = \int_{x_{i-1}}^{x_{i+1}} v_i(x) \frac{\partial D(x)}{\partial x} dx \qquad (4.4)$$

Note that in this case, the electric flux $D(x)$ is a scalar function. Let there be a centre node i which is surrounded by two segments A and B, defined by the interval (x_{i-1}, x_i) and (x_i, x_{i+1}) respectively, as shown Figure 4.2.

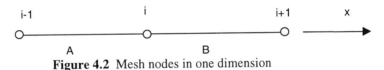

Figure 4.2 Mesh nodes in one dimension

Let the length of the segments and their dielectric permittivities in A and B be denoted by (h_A, h_B) and $(\varepsilon_A, \varepsilon_B)$ respectively. In this case, there are two "elements" (segments) surrounding the centre node i. The above integral can easily be integrated by parts so that

$$e_R = \int_{x_{i-1}}^{x_{i+1}} v_i \frac{\partial D}{\partial x} dx = [v_i D]_{x_{i-1}}^{x_{i+1}} - \int_{x_{i-1}}^{x_{i+1}} D \frac{\partial v_i}{\partial x} dx \qquad (4.5)$$

At this point the weighting function $v(x)$ needs to be defined. Let us assume that it is given by the hat function shown in Figure 4.3, which is equal to 1 at the centre node, and falls off linearly to 0 at the segment edges.

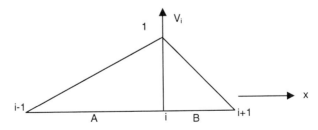

Figure 4.3 Hat function

Since

$$v_i(x_i) = 1 \qquad v_i(x_{i-1}) = 0 \qquad v_i(x_{i+1}) = 0$$

$$\frac{\partial v_i}{\partial x} = \frac{1}{h_A} \quad x_{i-1} \le x \le x_i \qquad \frac{\partial v_i}{\partial x} = -\frac{1}{h_B} \qquad x_i \le x \le x_{i+1} \tag{4.6}$$

so

$$\left[v_i D \right]_{x_{x-1}}^{x_{i+1}} = v_{i+1} D(x_{i+1}) - v_{i-1} D(x_{i-1}) = 0 \tag{4.7}$$

then the residual is simplified to

$$e_R = -\int_{x_{i-1}}^{x_i} \frac{D_A}{h_A} dx + \int_{x_i}^{x_{i+1}} \frac{D_B}{h_B} dx \tag{4.8}$$

The next step is specify the variation of $D(x)$. The most common form is to use a linear variation in electric potential, which of course is equivalent to assuming that the flux density is constant over each segment. This simplifies the residual to

$$e_R = -\frac{D_A}{h_A} \int_{x_{i-1}}^{x_i} dx + \frac{D_B}{h_B} \int_{x_i}^{x_{i+1}} dx \tag{4.9}$$

so

$$e_R = D_B - D_A \tag{4.10}$$

Making the residual zero is thus equivalent to specifying the divergence of D to be zero, in effect satisfying Gauss's law at the centre node

$$e_R = D_B - D_A = 0 = \frac{\partial D}{\partial x} = \nabla \cdot \mathbf{D} \tag{4.11}$$

This means that enforcing

$$\iint_R v(\nabla \cdot \mathbf{D}) dx dy = 0 \tag{4.12}$$

is equivalent to applying Gauss's Law over the domain R.

In this case, it is straightforward to write the electric flux directly in terms of the electric potential

$$D_A = -\varepsilon_A \frac{(V_i - V_{i-1})}{h_A} \tag{4.13}$$

$$D_B = -\varepsilon_B \frac{(V_{i+1} - V_i)}{h_B} \tag{4.14}$$

so the residual equation leads to the following nodal equation

$$-\varepsilon_A \frac{(V_i - V_{i-1})}{h_A} + \varepsilon_B \frac{(V_{i+1} - V_i)}{h_B} = 0 \tag{4.15}$$

which can be put in the following form

$$\varepsilon_A \frac{V_{i-1}}{h_A} + \varepsilon_B \frac{V_{i+1}}{h_B} - \left(\frac{\varepsilon_A}{h_A} + \frac{\varepsilon_B}{h_B}\right) V_i = 0 \tag{4.16}$$

This nodal equation is identical to the one from 5pt finite difference. But it is much more flexible since it naturally includes the effect of each segment having a different length and material property. In finite difference, a star of unequal arms needs to be considered and special fictitious points are introduced into each region and then eliminated. But here, the mesh size can be varied locally and regions of different material types can be specified by input data, requiring no modification to the nodal equation. In addition, a derivative boundary condition is quite easily specified. Consider the following boundary condition specified at the first mesh node

$$D(x_1) = -\varepsilon \frac{\partial V}{\partial x} = q \tag{4.17}$$

where h is the first segment length and ε is its permittivity. The residual is now

$$e_R = -\left[v_i \varepsilon \frac{\partial V}{\partial x}\right]_{x_1}^{x_2} + \int_{x_1}^{x_2} \varepsilon \frac{\partial V}{\partial x} \frac{\partial v_1}{\partial x} dx = 0 \tag{4.18}$$

which leads to

$$\frac{\varepsilon}{h}(V_2 - V_1) - q = 0 \tag{4.19}$$

This is the nodal equation for the first node, and is derived directly from input data.

Now it must be emphasised that for all the situations presented so far in this simple one dimensional case, it is fairly straightforward to confirm that the finite difference method will give exactly the same results as the finite element method. But in two and three dimensions, the advantages of the finite element method become clearer.

To apply derivative boundary conditions on curved boundaries requires considerable effort with the finite difference method, but in the finite element method, the problem of curved boundaries is essentially a problem of mesh generation, and boundary conditions and material type are specified as input data.

1.2   The variational approach

The weighted residual method formulates a partial differential equation in integral form. For a wide variety of differential equations, there is another way of formulating the equation in integral form, and it is called the variational approach. The variational approach leads to exactly the same nodal equations as the weighted residual approach. It often has the advantage of greater physical interpretation, but it is less general and only applies to equations which have the "self- adjoint" property and are "positive definite".

Consider the following partial differential equation over the domain R,

$$L(u) = f \tag{4.20}$$

If the operator L is self-adjoint, that is for two scalar functions, u and v,

$$\int_R L(u)v d\Omega = \int_R L(v)u dR \tag{4.21}$$

and positive definite, that is,

$$\int_R L(u)u dR > 0 \quad u \neq 0 \qquad \int_R L(u)u dR = 0 \quad u = 0 \tag{4.22}$$

then the solution can be found by minimising the following functional

$$F(u) = \frac{1}{2} \int_R \left(L(u)u - 2fu \right) dR \tag{4.23}$$

Consider a source-free electrostatic field (f=0), $L(u)=\nabla.(\varepsilon\nabla u)$, where $u(x,y,z)$ is the electrostatic potential and $\varepsilon(x,y,z)$ is the dielectric permittivity. $L(u)$ is self-adjoint, so the function $F(u)$ in this case is

$$F(u) = \frac{1}{2} \int_R u\nabla.(\varepsilon\nabla u) dxdydz \tag{4.24}$$

From Green's first scalar theorem (see Appendix 3)

$$F(u) = \iiint_R \left[u\nabla \cdot (\varepsilon\nabla u) \right] dR = -\iiint_R \varepsilon |\nabla u|^2 + \oiint_S u\varepsilon \frac{\partial u}{\partial n} dS \tag{4.25}$$

where S represents a surface which bounds the domain R. Consider the situation where the potential or its normal derivative on S is zero, the surface term is then omitted. The function F(u) under these conditions is equal in magnitude to the electrostatic energy W of the electric field and is given by the following well-known formula

$$W = |F(u)| = \frac{1}{2} \int \varepsilon |\nabla u|^2 \, dxdy \qquad (4.26)$$

The finite element then proceeds to minimise W with respect to u(x,y) at the mesh nodes in the domain.

The variational approach can be readily illustrated for the simple parallel plate capacitor example shown in Figure 4.4.

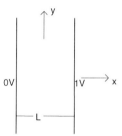

Figure 4.4 Parallel plate capacitor

The true potential solution varies linearly with x. Consider two false potential solutions, as shown in Figure 4.5.

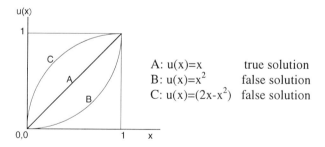

A: u(x)=x   true solution
B: u(x)=x^2 false solution
C: u(x)=(2x-x^2)   false solution

Figure 4.5 True and false solutions to the parallel plate capacitor problem

The energy of the system in this case is given by

$$W = \frac{1}{2} \int_0^1 \varepsilon \left(\frac{\partial u}{\partial x} \right)^2 dx \qquad (4.27)$$

Assuming $\varepsilon=1$, the energies of the three different solutions are given by

$$W_A=1/2, \quad W_B=2/3, \quad W_C=2/3 \qquad (4.28)$$

Clearly the energy is a minimum for the true solution. This principle can be extended to the potential at a node i: the potential u_i is varied until the energy at node i, W_i is minimised, so

$$\frac{\partial W_i}{\partial u_i} = 0 \qquad (4.29)$$

To show that the variational approach gives the same nodal equations as the weighted residual approach, consider the simple one-dimensional case of $u(x)$ as shown in Figure 4.2. The node i, is surrounded by segments A and B, and nodes i-1 and i+1.

In the variational approach, we first formulate W over each segment or element in the domain. Let W_A and W_B be the values of W over segments A and B respectively. So

$$W_A = \frac{1}{2} \int_{x_{i-1}}^{x_i} \varepsilon \left| \frac{\partial u}{\partial x} \right|^2 dx \qquad W_B = \frac{1}{2} \int_{x_i}^{x_{i+1}} \varepsilon \left| \frac{\partial u}{\partial x} \right|^2 dx \qquad (4.30)$$

If we assume that each segment has its own length and dielectric permittivity, h_A and ε_A for segment A, h_B and ε_B for segment B, and that the function $u(x)$ varies linearly with x, then

$$W_A = \frac{1}{2} \frac{\varepsilon_A}{h_A} \left(u_i - u_{i-1} \right)^2 \qquad W_B = \frac{1}{2} \frac{\varepsilon_B}{h_B} \left(u_{i+1} - u_i \right)^2 \qquad (4.31)$$

Here it is assumed that the dielectric permittivity is constant over each segment. Now we seek to minimise W with respect to u_i, since the node i is connected to node i-1 through segment A, and connected to node i+1 through segment B, then

$$\frac{\partial W}{\partial u_i} = \frac{\partial W_A}{\partial u_i} + \frac{\partial W_B}{\partial u_i} = \frac{\varepsilon_A}{h_A} \left(u_i - u_{i-1} \right) - \frac{\varepsilon_B}{h_B} \left(u_{i+1} - u_i \right) \qquad (4.32)$$

Thus putting the variation of W at node i to zero yields exactly the same residual equation found previously by the weighted residual method

$$u_{i-1} \left(\frac{\varepsilon_A}{h_A} \right) + u_{i+1} \left(\frac{\varepsilon_B}{h_B} \right) - u_i \left(\frac{\varepsilon_A}{h_A} + \frac{\varepsilon_B}{h_B} \right) = 0 \qquad (4.33)$$

In a two conductor system, the variational approach effectively minimises the capacitance between the conductors. Let ΔV be the voltage difference between the conductors and C be the capacitance of the system, then

$$W = \frac{1}{2}C(\Delta V)^2 \qquad (4.34)$$

So minimising W is equivalent to minimising C. The variational form of the finite element is particularly useful in calculating bulk parameters, such as the capacitance or the stored energy of an electrostatic system.

It is straightforward to formulate the variational approach in different coordinate systems. Consider the infinitely long rotationally symmetric structure shown in Figure 4.6

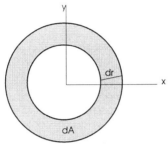

Figure 4.6 An infinitely long rotationally symmetric structure

The energy dW associated with an element dr extending in the radial direction is

$$dW = \frac{1}{2}\varepsilon\left[\left(\frac{\partial u}{\partial r}\right)^2\right](dA) \qquad (4.35)$$

where the area dA of the element is clearly given by

$$dA = 2\pi r(dr) \qquad (4.36)$$

so

$$W = \pi \int \varepsilon\left(\frac{\partial u}{\partial r}\right)^2 r dr \qquad (4.37)$$

The variational method then proceeds in exactly the same way as before, where the energy W is found for a node and its surrounding elements, and then differentiated with respect to the unknown potential at that node. In spherical coordinates with symmetry in both angular directions, the energy W of the system is given by

FINITE ELEMENT CONCEPTS

$$W = 2\pi \int \varepsilon \left(\frac{\partial u}{\partial r}\right)^2 r^2 dr \qquad (4.38)$$

2. THE VARIATIONAL METHOD IN TWO DIMENSIONS

2.1 Square Elements

Consider the nodal layout with square elements shown in Figure 4.7

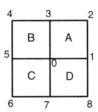

Figure 4.7 Square element nodal layout

Assume that there is a different permittivity in each quadrant, ε_A, ε_B, ε_C, and ε_D. Let the spacing between nodes be h. The finite element method always involves making an assumption about how the potential varies within an element. The following analysis assumes that the electric field components are constant over each quadrant region, which means that the potential is assumed to vary linearly.

Over region A, assume $\quad \left(\dfrac{\partial u}{\partial x}\right)_A = \left(\dfrac{u_1 - u_0}{h}\right) \quad \left(\dfrac{\partial u}{\partial y}\right)_A = \left(\dfrac{u_3 - u_0}{h}\right)$

Over region B, assume $\quad \left(\dfrac{\partial u}{\partial x}\right)_B = \left(\dfrac{u_0 - u_5}{h}\right) \quad \left(\dfrac{\partial u}{\partial y}\right)_B = \left(\dfrac{u_3 - u_0}{h}\right)$

Over region C, assume $\quad \left(\dfrac{\partial u}{\partial x}\right)_C = \left(\dfrac{u_0 - u_5}{h}\right) \quad \left(\dfrac{\partial u}{\partial y}\right)_C = \left(\dfrac{u_0 - u_7}{h}\right)$

Over region D, assume $\quad \left(\dfrac{\partial u}{\partial x}\right)_D = \left(\dfrac{u_1 - u_0}{h}\right) \quad \left(\dfrac{\partial u}{\partial y}\right)_D = \left(\dfrac{u_0 - u_7}{h}\right) \qquad (4.39)$

The electrostatic energy is

$$W = \frac{1}{2} \int \varepsilon \left[\left(\frac{\partial u}{\partial x}\right)^2 + \left(\frac{\partial u}{\partial y}\right)^2\right] dx dy \qquad (4.40)$$

Since the gradients of the potential are constant over the element, and the area of each element is given by h^2, then the energy for each element is given by

$$W_A = \frac{\varepsilon_A}{2}\left[(u_1 - u_0)^2 + (u_3 - u_0)^2\right] \quad W_B = \frac{\varepsilon_B}{2}\left[(u_0 - u_5)^2 + (u_3 - u_0)^2\right]$$

$$W_C = \frac{\varepsilon_C}{2}\left[(u_0 - u_5)^2 + (u_0 - u_7)^2\right] \quad W_D = \frac{\varepsilon_D}{2}\left[(u_1 - u_0)^2 + (u_0 - u_7)^2\right] \quad (4.41)$$

The variational principle requires that the total energy around node 0 must have a minimum, so

$$\frac{\partial W}{\partial u_0} = \frac{\partial(W_A + W_B + W_C + W_D)}{\partial u_0} = 0 \tag{4.42}$$

Differentiating each element with respect to the centre node potential, the nodal equation becomes

$$2u_0(\varepsilon_A + \varepsilon_B + \varepsilon_C + \varepsilon_D) - u_1(\varepsilon_A + \varepsilon_D) - u_3(\varepsilon_A + \varepsilon_B) - u_5(\varepsilon_B + \varepsilon_C) - u_7(\varepsilon_C + \varepsilon_D) = 0$$

$$(4.43)$$

The corresponding pictorial star is

$$\left(\begin{array}{ccc} & -(\varepsilon_A + \varepsilon_B) & \\ -(\varepsilon_B + \varepsilon_C) & 2(\varepsilon_A + \varepsilon_B + \varepsilon_C + \varepsilon_D) & -(\varepsilon_A + \varepsilon_D) \\ & -(\varepsilon_C + \varepsilon_D) & \end{array}\right) u = 0 \tag{4.44}$$

The permittivity of the regions on either side of each arm is involved, and the centre coefficient is obtained by the sum of all the region permittivities.

The above pictorial star can be directly applied to problems which have regular mesh spacing. Consider the two-dimensional box shown in Figure 4.8. A dielectric slab of relative permittivity ε_R partially fills the box.

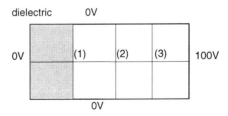

Figure 4.8 A partially filled box

By letting $S=\varepsilon_R$, the first-order finite element approximation to the problem leads to the following matrix equation

$$\begin{pmatrix} -2(1+S) & 1 & 0 \\ 1 & -4 & 1 \\ 0 & 1 & -4 \end{pmatrix} \begin{pmatrix} u_1 \\ u_2 \\ u_3 \end{pmatrix} = \begin{pmatrix} 0 \\ 0 \\ -100 \end{pmatrix} \qquad (4.45)$$

The quadrant region of another problem is illustrated in Figure 4.9. Here a dielectric slab is placed at the centre of a square transmission line.

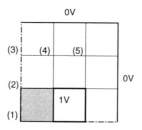

Figure 4.9 Partially filled square transmission line

Let the shaded area be a dielectric with permittivity ε_R. The resulting matrix equation is given by

$$\begin{pmatrix} 2\varepsilon_R & -\varepsilon_R & 0 & 0 & 0 \\ -\varepsilon_R & 2(1+\varepsilon_R) & -1 & 0 & 0 \\ 0 & -1 & 4 & -2 & 0 \\ 0 & 0 & -1 & 4 & -1 \\ 0 & 0 & 0 & -1 & 4 \end{pmatrix} \begin{pmatrix} u_1 \\ u_2 \\ u_3 \\ u_4 \\ u_5 \end{pmatrix} = \begin{pmatrix} \varepsilon_R \\ 1+\varepsilon_R \\ 0 \\ 1 \\ 1 \end{pmatrix} \qquad (4.46)$$

2.2   Rectangular elements

Consider rectangle elements, as depicted in Figure 4.10.

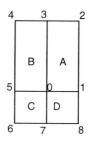

Figure 4.10 Rectangle element layout

Assume that the spacing from node 0 to node 1 is h_1, to node 3 is h_3, to node 5 is h_5, and to node 7 is h_7. Using the same approach as previously used, where the electric field within each element is assumed to be constant, it is straightforward to show that the variational approach leads to the following nodal equation

$$u_0\left[\varepsilon_A\left(\frac{h_1}{h_3}+\frac{h_3}{h_1}\right)+\varepsilon_B\left(\frac{h_3}{h_5}+\frac{h_5}{h_3}\right)+\varepsilon_C\left(\frac{h_5}{h_7}+\frac{h_7}{h_5}\right)+\varepsilon_D\left(\frac{h_1}{h_7}+\frac{h_7}{h_1}\right)\right]$$

$$-u_1\left(\frac{h_3}{h_1}\varepsilon_A+\frac{h_7}{h_1}\varepsilon_D\right)-u_3\left(\frac{h_1}{h_3}\varepsilon_A+\frac{h_5}{h_3}\varepsilon_B\right)-u_5\left(\frac{h_3}{h_5}\varepsilon_B+\frac{h_7}{h_5}\varepsilon_C\right)-u_7\left(\frac{h_5}{h_7}\varepsilon_C+\frac{h_1}{h_7}\varepsilon_D\right)=0 \qquad (4.47)$$

The same nodal equation is obtained by the use of an asymmetric 5 pt finite difference star.

2.3   Right-angle Triangle elements

Consider the triangle element nodal layout depicted 4.11

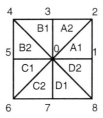

Figure 4.11 Right-angle triangle element nodal layout

Assume that $\varepsilon_{A1}\neq\varepsilon_{A2}$, $\varepsilon_{B1}\neq\varepsilon_{B2}$, $\varepsilon_{C1}\neq\varepsilon_{C2}$, $\varepsilon_{D1}\neq\varepsilon_{D2}$ and that the nodes are evenly spaced by a distance h.

We start by making an approximation for the variation of the potential within each element. A linear variation is equivalent to assuming that the electric field strength is constant over each element. The gradient of the potential across each right triangle is constant and given in terms of the potentials at the vertices

$$\left(\frac{\partial u}{\partial x}\right)_{A1}=\frac{u_1-u_0}{h} \quad \left(\frac{\partial u}{\partial y}\right)_{A1}=\frac{u_2-u_1}{h} \qquad \left(\frac{\partial u}{\partial x}\right)_{A2}=\frac{u_2-u_3}{h} \quad \left(\frac{\partial u}{\partial y}\right)_{A2}=\frac{u_3-u_0}{h}$$

$$\left(\frac{\partial u}{\partial x}\right)_{B1}=\frac{u_3-u_4}{h} \quad \left(\frac{\partial u}{\partial y}\right)_{B1}=\frac{u_3-u_0}{h} \qquad \left(\frac{\partial u}{\partial x}\right)_{B2}=\frac{u_0-u_5}{h} \quad \left(\frac{\partial u}{\partial y}\right)_{B2}=\frac{u_4-u_5}{h}$$

etc. $\hspace{12cm}$ (4.48)

The variational principle states that

$$\frac{\partial W}{\partial u_0} = \frac{\partial \left(W_{A1} + W_{A2} + W_{B1} + W_{B2} + W_{C1} + W_{C2} + W_{D1} + W_{D2} \right)}{\partial u_0} = 0 \qquad (4.49)$$

This leads to the following picture operator

$$\left(\begin{array}{ccc} & -\left(\varepsilon_{A2} + \varepsilon_{B1}\right) & \\ -\left(\varepsilon_{B2} + \varepsilon_{C1}\right) & \left(\varepsilon_{A1} + \varepsilon_{A2} + \varepsilon_{B1} + \varepsilon_{B2} + \varepsilon_{C1} + \varepsilon_{C2} + \varepsilon_{D1} + \varepsilon_{D2}\right) & -\left(\varepsilon_{A1} + \varepsilon_{D2}\right) \\ & -\left(\varepsilon_{C2} + \varepsilon_{D1}\right) & \end{array} \right) u = 0$$

$$(4.50)$$

This pictorial star is equivalent to the one derived for square elements having $\varepsilon_{A1} = \varepsilon_{A2} = \varepsilon_A$, $\varepsilon_{B1} = \varepsilon_{B2} = \varepsilon_B$, $\varepsilon_{C1} = \varepsilon_{C2} = \varepsilon_C$, and $\varepsilon_{D1} = \varepsilon_{D2} = \varepsilon_D$. Moreover, it is possible to show that the nodal equations for pictorial stars depicted in Figure 4.12a and 4.12b are identical.

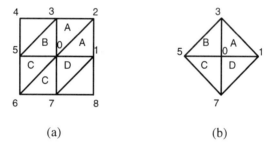

(a)   (b)

Figure 4.12 Equivalent right-angle triangle element nodal layouts

Using this kind of property, a wide variety of different problems can be solved by simply applying the pictorial star. Consider the right-angle triangle element layout depicted in Figure 4.13

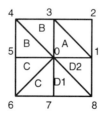

Figure 4.13 Right-angle triangle example layout

The picture operator for this case is given by

$$\left(\begin{matrix} & -\left(\varepsilon_A + \varepsilon_B\right) & \\ -\left(\varepsilon_B + \varepsilon_C\right) & \left(2\varepsilon_A + 2\varepsilon_B + 2\varepsilon_C + \varepsilon_{D1} + \varepsilon_{D2}\right) & -\left(\varepsilon_A + \varepsilon_{D2}\right) \\ & -\left(\varepsilon_C + \varepsilon_{D1}\right) & \end{matrix} \right) u = 0 \tag{4.51}$$

This pictorial star can be applied to solve problems such as the one depicted in Figure 4.14.

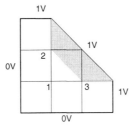

Figure 4.14 Partially filled wave guide with oblong shaped dielectric slab

Let the shaded area be a dielectric with permittivity ε_R. The matrix equation for the potential nodes 1, 2 and 3 from first-order finite elements is given by

$$\begin{pmatrix} 4 & -1 & -1 \\ -1 & \frac{1}{2}\left(3\varepsilon_R + 5\right) & 0 \\ -1 & 0 & \frac{1}{2}\left(3\varepsilon_R + 5\right) \end{pmatrix} \begin{pmatrix} u_1 \\ u_2 \\ u_3 \end{pmatrix} = \begin{pmatrix} 0 \\ \dfrac{3\varepsilon_R}{2} + \dfrac{1}{2} \\ \dfrac{3\varepsilon_R}{2} + \dfrac{1}{2} \end{pmatrix} \tag{4.52}$$

3   FIRST-ORDER SHAPE FUNCTIONS

In the finite element method, the whole domain of interest is divided into small elements. The most popular element shape in two dimensions is the triangle. One of the flexible features of the finite element method is that a variety of different element shapes can be chosen. A typical finite element mesh is shown in Figure 4.15.

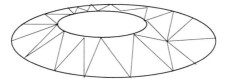

Figure 4.15 A Triangle Element mesh

Triangle elements are popular since they can be the basic building block for a wide variety of shapes. The form of the function to be calculated over each element must be specified. The most widely used variation is the linear one, but higher-order variations can in many cases be advantageous. A general method of expressing the variation of a function over elements with different geometries is based upon the use of "shape functions".

3.1   Shape functions in one dimension

Consider the linear variation of the function $u(x)$ over a segment of length h between nodes denoted by 1 and 2, as shown in Figure 4.16.

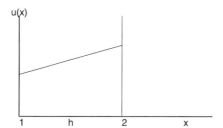

Figure 4.16 Linear variation over an element in one dimension

Let the potential $u(x)$ be of the following form

$$u(x)=a+bx \qquad (4.53)$$

where a and b are constants. Let the element interval be (x_1, x_2), and (u_1, u_2) be the potential values at nodes 1 and 2. Shape functions relate the function $u(x)$ to its values at the mesh nodes. In this case there are obviously two functions $N_1(x)$ and $N_2(x)$, so that

$$u(x) = N_1(x)u_1 + N_2(x)u_2 \qquad (4.54)$$

By inspection, the requirements for the shape functions are

$$N_1(x_1) = 1 \qquad N_1(x_2) = 0 \qquad N_2(x_1) = 0 \qquad N_2(x_2) = 1 \qquad (4.55)$$

By substituting for the values of x at nodes 1 and 2, the constants a and b can easily be found. The shape functions in this simple one dimensional case are obviously given by

$$N_1(x) = \left(1 - \frac{x}{h}\right) \qquad N_2(x) = \frac{x}{h} \qquad\qquad (4.56)$$

These shape functions can be thought of as "hat" functions, as shown in Figure 4.17. For every node in the domain, there are a set of shape functions which are equal to unity at the coordinates of the node, and zero at all the other nodes.

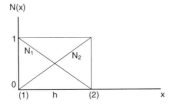

Figure 4.17 Shape functions in one dimension

The general requirement for the shape functions are as follows

$$\begin{aligned} N_i(x_j) &= 1 \quad j = i \\ &= 0 \quad j \neq i \end{aligned} \qquad\qquad (4.57)$$

The shape function is equal to unity at node i, and zero at all the other nodes

3.2   Shape functions in two dimensions

Consider the triangle element shown in Figure 4.18. The corners are numbered in an anti-clockwise direction.

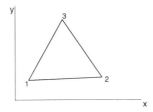

Figure 4.18 Triangle element

Restricting the discussion to linear function variations across the triangle

$$u(x, y) = f + gx + hy \qquad\qquad (4.58)$$

where f, g and h are constants. In this case there are three shape functions, one for each node, so that

$$u(x, y) = N_1(x, y)u_1 + N_2(x, y)u_2 + N_3(x, y)u_3 \qquad (4.59)$$

One way of solving for f, g and h is to substitute in the coordinates (x,y) at the corners, so

$$\begin{pmatrix} u_1 \\ u_2 \\ u_3 \end{pmatrix} = \begin{pmatrix} 1 & x_1 & y_1 \\ 1 & x_2 & y_2 \\ 1 & x_3 & y_3 \end{pmatrix} \begin{pmatrix} f \\ g \\ h \end{pmatrix} \qquad (4.60)$$

Let the matrix [A] be defined as

$$A = \begin{pmatrix} 1 & x_1 & y_1 \\ 1 & x_2 & y_2 \\ 1 & x_3 & y_3 \end{pmatrix} \qquad (4.61)$$

and let its inverse be defined by

$$A^{-1} = (\det A)^{-1} \begin{pmatrix} a_1 & a_2 & a_3 \\ b_1 & b_2 & b_3 \\ c_1 & c_2 & c_3 \end{pmatrix} \qquad (4.62)$$

where the $a_1....c_3$ coefficients can easily be found analytically. The parameters f, g, and h are then solved by

$$\begin{pmatrix} f \\ g \\ h \end{pmatrix} = (\det A)^{-1} \begin{pmatrix} a_1 & a_2 & a_3 \\ b_1 & b_2 & b_3 \\ c_1 & c_2 & c_3 \end{pmatrix} \begin{pmatrix} u_1 \\ u_2 \\ u_3 \end{pmatrix} \qquad (4.63)$$

So the shape functions can be determined from

$$u = \sum_{j=1}^{3} N_j u_j = f + gx + hy \qquad (4.64)$$

and they are

$$N_j(x,y) = \frac{1}{(\det A)}\left(a_j + b_j x + c_j y\right) \qquad j=1,2,3$$

$$
\begin{aligned}
a_1 &= x_2 y_3 - x_3 y_2 \quad a_2 = x_3 y_1 - x_1 y_3 \quad a_3 = x_1 y_2 - x_2 y_1 \\
b_1 &= y_2 - y_3 \qquad\quad b_2 = y_3 - y_1 \quad b_3 = y_1 - y_2 \\
c_1 &= x_3 - x_2 \qquad\quad c_2 = x_1 - x_3 \quad c_3 = x_2 - x_1 \\
\det A &= c_2 b_1 - c_1 b_2
\end{aligned}
\tag{4.65}
$$

The shape functions are thus purely geometrical and depend on the shape of the element. Note that the gradients of the shape function in x and y are given by

$$\frac{\partial N_j}{\partial x} = \frac{1}{(\det A)} b_j \qquad \frac{\partial N_j}{\partial y} = \frac{1}{(\det A)} c_j \tag{4.66}$$

It can be shown that the determinant of the matrix [A] is equal to twice the area of the triangle.

3.3   Example of a right angle triangle

Consider a right angle triangle with side lengths p by q, as shown in Figure 4.19

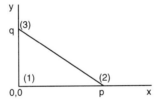

Figure 4.19 Right-angle triangle element in local element numbering

Let the local node numbering be (1), (2) and (3) as shown. The shape functions can then be calculated from equation (4.65) since the coordinates of each point are known

Node	x	y	a	b	c
1	0	0	pq	-q	-p
2	p	0	0	q	0
3	0	q	0 0	p	

$\det A = pq = 2 \times$ Area, and

$$N_1 = 1 - \frac{x}{p} - \frac{y}{q} \qquad N_2 = \frac{x}{p} \qquad N_3 = \frac{y}{q} \tag{4.67}$$

Note that the shape functions sum up to unity. Note also that $N_1(0,0)=1$, $N_2(0,0)=0$, $N_3(0,0)=0$, confirming that

$$\begin{aligned} N_i(x_j) &= 1 \qquad j = i \\ &= 0 \qquad j \neq i \qquad\qquad i, j = 1,2,3 \end{aligned} \tag{4.68}$$

The shape functions are interpolation functions and give an estimate of the point u at the point (x,y) in terms of the potentials at the nodes. They give an average value of potential for an average value of the coordinates

$$\overline{x} = \frac{1}{3}(x_1 + x_2 + x_3) \quad \overline{y} = \frac{1}{3}(y_1 + y_2 + y_3) \quad \overline{u} = \frac{1}{3}(u_1 + u_2 + u_3) \tag{4.69}$$

At the mid-side point of any side of the triangle, the potential is equal to the average of the respective corner potentials. The shape function gradients are easily found to be

$$\frac{\partial N_1}{\partial x} = -\frac{1}{p} \quad \frac{\partial N_1}{\partial y} = -\frac{1}{q} \quad \frac{\partial N_2}{\partial x} = \frac{1}{p} \quad \frac{\partial N_2}{\partial y} = 0 \quad \frac{\partial N_3}{\partial x} = 0 \quad \frac{\partial N_3}{\partial y} = \frac{1}{q} \tag{4.70}$$

The gradients of the shape functions can be interpreted as being normalised derivatives. Consider for instance applying a unit voltage to node 1, and 0 volts to nodes 2 and 3, then the gradients of the potential along the (1) - (2) edge and (1)-(3) give

$$\frac{\partial N_1}{\partial x} = \frac{u_2 - u_1}{p} = \frac{0-1}{p} = -\frac{1}{p} \qquad \frac{\partial N_1}{\partial y} = \frac{u_3 - u_1}{q} = \frac{0-1}{q} = -\frac{1}{q} \tag{4.71}$$

Now apply a unit voltage to node 2, and zero volts to nodes 1 and 3, so the gradients of the shape function 2 are

$$\frac{\partial N_2}{\partial x} = \frac{u_2 - u_1}{p} = \frac{1-0}{p} = \frac{1}{p} \qquad \frac{\partial N_2}{\partial y} = \frac{u_3 - u_1}{q} = \frac{0}{q} = 0 \tag{4.72}$$

Now apply a unit voltage to node 3, and zero volts to nodes 1 and 2, so the gradients of the shape function 3 are

$$\frac{\partial N_3}{\partial x} = \frac{u_2 - u_1}{p} = \frac{0}{p} = 0 \qquad \frac{\partial N_3}{\partial y} = \frac{u_3 - u_1}{q} = \frac{1-0}{q} = \frac{1}{q} \tag{4.73}$$

We can derive the gradient of the potential at any point (x,y) in the triangle by

$$\frac{\partial u}{\partial x} = \frac{\partial N_1}{\partial x} u_1 + \frac{\partial N_2}{\partial x} u_2 + \frac{\partial N_3}{\partial x} u_3 = \frac{1}{p}\left(u_2 - u_1\right)$$

$$\frac{\partial u}{\partial y} = \frac{\partial N_1}{\partial y} u_1 + \frac{\partial N_2}{\partial y} u_2 + \frac{\partial N_3}{\partial y} u_3 = \frac{1}{q}\left(u_3 - u_1\right) \qquad (4.74)$$

Clearly the gradients of the potential are constant over the triangle and they are equal to the gradient of the potential along the edges of the triangle.

Consider relabelling the right angle triangle, as shown in Figure 4.20

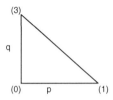

Figure 4.20 Right-angle triangle element in local nodal numbering

The gradients become

$$\frac{\partial N_0}{\partial x} = -\frac{1}{p} \quad \frac{\partial N_0}{\partial y} = -\frac{1}{q} \qquad \frac{\partial u}{\partial x} = \frac{1}{p}\left(u_1 - u_0\right) \quad \frac{\partial u}{\partial y} = \frac{1}{q}\left(u_3 - u_0\right) \qquad (4.75)$$

The term $\nabla N_0.\nabla u$ is given by

$$\nabla N_0.\nabla u = \left(\frac{\partial N_0}{\partial x}\right)\left(\frac{\partial u}{\partial x}\right) + \left(\frac{\partial N_0}{\partial y}\right)\left(\frac{\partial u}{\partial y}\right) = -\frac{1}{p^2}\left(u_1 - u_0\right) - \frac{1}{q^2}\left(u_3 - u_0\right) \qquad (4.76)$$

Consider the alternative subdivision into triangles A and B with nodes (0), (1), (2) and (3) as shown in Figure 4.21

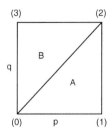

Figure 4.21 Alternative right-angle triangle element subdivision

For triangle A

$$\frac{\partial u}{\partial x} = \frac{1}{p}(u_1 - u_0) \qquad \frac{\partial u}{\partial y} = \frac{1}{q}(u_2 - u_1) \qquad \frac{\partial N_0}{\partial x} = -\frac{1}{p} \qquad \frac{\partial N_0}{\partial y} = 0 \qquad (4.77)$$

so

$$(\nabla N_0.\nabla u)_A = \left(\frac{\partial N_0}{\partial x}\right)\left(\frac{\partial u}{\partial x}\right) + \left(\frac{\partial N_0}{\partial y}\right)\left(\frac{\partial u}{\partial y}\right) = -\frac{1}{p^2}(u_1 - u_0) \qquad (4.78)$$

and for triangle B

$$\frac{\partial u}{\partial x} = \frac{1}{p}(u_2 - u_3) \qquad \frac{\partial u}{\partial y} = \frac{1}{q}(u_3 - u_0) \qquad \frac{\partial N_0}{\partial x} = 0 \qquad \frac{\partial N_0}{\partial y} = -\frac{1}{q} \qquad (4.79)$$

so

$$(\nabla N_0.\nabla u)_B = \left(\frac{\partial N_0}{\partial x}\right)\left(\frac{\partial u}{\partial x}\right) + \left(\frac{\partial N_0}{\partial y}\right)\left(\frac{\partial u}{\partial y}\right) = -\frac{1}{q^2}(u_3 - u_0) \qquad (4.80)$$

So for the square element as a whole, $\nabla N_0.\nabla u$ is equal to the right angle triangle of the previous case

$$(\nabla N_0.\nabla u)_A + (\nabla N_0.\nabla u)_B = -\frac{1}{p^2}(u_1 - u_0) - \frac{1}{q^2}(u_3 - u_0) \qquad (4.81)$$

3.4   Quadrilateral element shape functions

Apart from the triangle, another popular element for two dimensional problems is the quadrilateral element. To begin with, consider the special case of a rectangular element with sides of length a and b, as shown in Figure 4.22 below.

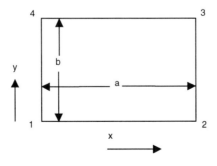

Figure 4.22 Rectangle element

Let the (x,y) coordinates for the nodes 1,2,3 and 4 be (0,0), (a,0), (a,b) and (0,b) respectively. Here the shape elements are simply given by

$$N_1(x,y) = \frac{1}{ab}(a-x)(b-y) \qquad N_2(x,y) = \frac{x}{ab}(b-y)$$

$$N_3(x,y) = \frac{yx}{ab} \qquad\qquad N_4(x,y) = \frac{y}{ab}(a-x)$$

(4.82)

The reader can verify that $N_i(x_jy_j)$ is equal to unity for i=j and zero for i≠j.

Unlike the first-order triangle element, interpolation for the potential across the element includes the cross-term xy, so

$$u(x,y) = \sum_{j=1}^{4} N_j(x,y)u_j = u_1 + \frac{x}{a}(u_2 - u_1) + \frac{y}{b}(u_4 - u_1) + \frac{xy}{ab}(u_1 - u_2 + u_3 - u_4)$$

(4.83)

This means that the gradients of the shape function will not be constant over the element. They will vary linearly with x and y, and are given by

$$\frac{\partial u}{\partial x} = \frac{1}{ab}\left[-u_1(b-y) + u_2(b-y) + u_3 y - u_4 y\right]$$

$$\frac{\partial u}{\partial y} = \frac{1}{ab}\left[-u_1(a-x) - u_2 x + u_3 x + u_4(a-x)\right]$$

(4.84)

For a general first-order quadrilateral element, the shape functions are best defined in normalised coordinates, (u,v), so that

$$N_1(u,v) = \frac{1}{4}(1-u)(1-v) \qquad N_2(u,v) = \frac{1}{4}(1+u)(1-v)$$

$$N_3(u,v) = \frac{1}{4}(1+u)(1+v) \qquad N_4(u,v) = \frac{1}{4}(1-u)(1+v)$$

(4.85)

where $-1 \le u \le 1$ and $-1 \le v \le 1$. The shape functions map the quadrilateral element on to a square region in normalised space, as shown in Figure 4.23.

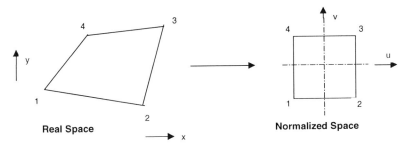

Figure 4.23 Quadrilateral element shape function mapping

The potential u, x and y in the element are expressed in terms of their values at the corner nodes

$$u(x, y) = \sum_{j=1}^{4} N_j u_j \qquad x = \sum_{j=1}^{4} N_j x_j \qquad y = \sum_{j=1}^{4} N_j y_j \qquad (4.86)$$

To derive the derivatives of the potential or shape functions, the Jacobian matrix of the transformation is needed. Consider the problem of finding the shape function derivatives. They can be expressed as derivatives in the (u,v) variables by

$$\frac{\partial N_i}{\partial u} = \frac{\partial N_i}{\partial x}\frac{\partial x}{\partial u} + \frac{\partial N_i}{\partial y}\frac{\partial y}{\partial u} \qquad \frac{\partial N_i}{\partial v} = \frac{\partial N_i}{\partial x}\frac{\partial x}{\partial v} + \frac{\partial N_i}{\partial y}\frac{\partial y}{\partial v} \qquad (4.87)$$

So the shape function derivatives with respect to x and y are found by

$$\begin{pmatrix} \dfrac{\partial N_i}{\partial x} \\[2mm] \dfrac{\partial N_i}{\partial y} \end{pmatrix} = J^{-1} \begin{pmatrix} \dfrac{\partial N_i}{\partial u} \\[2mm] \dfrac{\partial N_i}{\partial v} \end{pmatrix} \qquad J = \begin{pmatrix} \dfrac{\partial x}{\partial u} & \dfrac{\partial y}{\partial u} \\[2mm] \dfrac{\partial x}{\partial v} & \dfrac{\partial y}{\partial v} \end{pmatrix} \qquad (4.88)$$

Once these derivatives are found, the derivatives of the potential can then be found from

$$\frac{\partial u}{\partial x} = \sum_{j=1}^{4} \frac{\partial N_j}{\partial x} u_j \qquad \frac{\partial u}{\partial y} = \sum_{j=1}^{4} \frac{\partial N_j}{\partial y} u_j \qquad (4.89)$$

These derivatives are clearly dependent on the (x,y) coordinates of the element nodes.

4.   The Galerkin Method

In two dimensions, the weighted residual of an equation L(u)=0 over the domain Ω is given by

$$\iint_{\Omega} vL(u)dxdy = 0 \qquad (4.90)$$

Now for an electrostatic layout which has no free charge this becomes

$$\iint_{\Omega} v\nabla \cdot (\varepsilon\nabla u)dxdy = 0 \qquad (4.91)$$

To expand this integral, we use Green's first scalar theorem given in Appendix 3. So

$$\iint_{\Omega} v\nabla \cdot (\varepsilon\nabla u)dxdy = -\iint_{\Omega} \varepsilon\nabla v \cdot (\nabla u)dxdy + \oint \varepsilon v \frac{\partial u}{\partial n}dS \qquad (4.92)$$

where S refers to the surface of the outer boundary enclosing the domain.

Note the term integrated along the outer boundary is zero if the normal derivative of u is zero. If the potential on it is given explicitly, then it is also omitted since v=0 on S for the nodes next to the boundary. The node on the boundary itself is taken into account as an explicit potential value. For the moment we will assume that this term is zero. We consider the weighted average of the residual at node i and enforce it to be zero

$$\iint_{\Omega} \varepsilon\nabla v_{i} \cdot (\nabla u)dxdy = 0 \qquad (4.93)$$

Now we look for a suitable weighting function v_i. Previously we used a hat function, which equals to unity at the node i, and which falls to zero at the surrounding nodes. A natural choice for the weighting function which has precisely this property is the finite element shape function. Putting $v_i(x,y)=N_i(x,y)$ is called the Galerkin method, and it is a convenient form of the weighted residual method for finite elements. It effectively localizes the residual around each point in the same way that the basis function itself is localized.

The residual equation becomes

$$\iint_{\Omega} \varepsilon\nabla N_{i} \cdot (\nabla u)dxdy = \iint_{\Omega} \varepsilon\left[\left(\frac{\partial N_i}{\partial x}\right)\left(\frac{\partial u}{\partial x}\right) + \left(\frac{\partial N_i}{\partial y}\right)\left(\frac{\partial u}{\partial y}\right)\right]dxdy = 0 \qquad (4.94)$$

The next step is to formulate the residual equation for a node and its nearest neighbors. Consider the right-angle triangle subdivision shown in Figure 4.24.

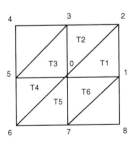

Figure 4.24 Right-angle triangle subdivision

Let h be the spacing between the nodes. This is only one possible scheme to divide the region surrounding a node into triangles (T1 to T6), there are other possible ways. The residual equation is obviously localized by the shape functions of these triangles. The centre node is given by 0.

Let R_{T1} be the contribution from element T1

$$R_{T1} = \iint_{T1} \varepsilon \left[\left(\frac{\partial N_0}{\partial x} \right) \left(\frac{\partial u}{\partial x} \right) + \left(\frac{\partial N_0}{\partial y} \right) \left(\frac{\partial u}{\partial y} \right) \right] dxdy \qquad (4.95)$$

and so on for each triangle. Based upon the previous result of 1st order shape functions for right-angle triangles (equations 4.76 and 4.81), it follows that

$$(\nabla N_0 . \nabla u)_{T1} = -\frac{1}{h^2}(u_1 - u_0) \qquad\qquad (\nabla N_0 . \nabla u)_{T3} = -\frac{1}{h^2}(u_5 - u_0) - \frac{1}{h^2}(u_3 - u_0)$$

$$(\nabla N_0 . \nabla u)_{T2} = -\frac{1}{h^2}(u_3 - u_0) \qquad\qquad (\nabla N_0 . \nabla u)_{T5} = -\frac{1}{h^2}(u_7 - u_0)$$

$$(\nabla N_0 . \nabla u)_{T4} = -\frac{1}{h^2}(u_5 - u_0) \qquad\qquad (\nabla N_0 . \nabla u)_{T6} = -\frac{1}{h^2}(u_7 - u_0) - \frac{1}{h^2}(u_1 - u_0)$$

$$(4.96)$$

Note the difference in contribution where the centre node lies on a right angle element corner. Let the relative permittivity of each triangle be $\varepsilon_{T1}, \varepsilon_{T2}, ... \varepsilon_{T6}$ and the permittivity be different for each quadrant, so $\varepsilon_{T1} = \varepsilon_{T2} = \varepsilon_A$, $\varepsilon_{T3} = \varepsilon_B$, $\varepsilon_{T4} = \varepsilon_{T5} = \varepsilon_C$, $\varepsilon_{T6} = \varepsilon_D$, as shown in Figure 4.25.

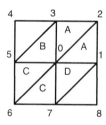

Figure 4.25 Right-angle triangle subdivision into regions A, B, C and D

Since the area of each triangle is $h^2/2$ then

$$R_{T1} = \frac{\varepsilon_A}{2}(u_0 - u_1) \qquad R_{T2} = \frac{\varepsilon_A}{2}(u_0 - u_3) \qquad R_{T3} = \frac{\varepsilon_B}{2}(2u_0 - u_5 - u_3)$$

$$R_{T4} = \frac{\varepsilon_C}{2}(u_0 - u_5) \qquad R_{T5} = \frac{\varepsilon_C}{2}(u_0 - u_7) \qquad R_{T6} = \frac{\varepsilon_D}{2}(2u_0 - u_7 - u_1)$$

$$(4.97)$$

The nodal equation is

$$R_{T1} + R_{T2} + ...R_{T6} = 0 \qquad (4.98)$$

So substituting in the contribution from each triangle gives the following nodal equation

$$2u_0\left(\varepsilon_A + \varepsilon_B + \varepsilon_C + \varepsilon_D\right) - u_1\left(\varepsilon_A + \varepsilon_D\right) - u_3\left(\varepsilon_A + \varepsilon_B\right) - u_5\left(\varepsilon_B + \varepsilon_C\right) - u_7\left(\varepsilon_C + \varepsilon_D\right) = 0$$

(4.99)

Note that this equation is identical to the one produced by the 5 pt finite difference star for constant permittivity and is identical to the results derived by the variational method in section 2 of this chapter.

The Galerkin approach can of course apply to any element type, so long as its shape functions can be written down. Keeping the same regions, A, B, C and D, consider first-order square elements having sides of length h, where the shape functions are given by

$$N_1(x, y) = \frac{1}{h^2}(h - x)(h - y) \qquad N_2(x, y) = \frac{x}{h^2}(h - y)$$

(4.100)

$$N_3(x, y) = \frac{yx}{h^2} \qquad N_4(x, y) = \frac{y}{h^2}(h - x)$$

It is straightforward to show in this case that the picture star is given by

$$\frac{2}{3}\begin{pmatrix} -\dfrac{\varepsilon_B}{2} & -\dfrac{1}{4}\left(\varepsilon_A + \varepsilon_B\right) & -\dfrac{\varepsilon_A}{2} \\[2mm] -\dfrac{1}{4}\left(\varepsilon_C + \varepsilon_B\right) & \left(\varepsilon_A + \varepsilon_B + \varepsilon_C + \varepsilon_D\right) & -\dfrac{1}{4}\left(\varepsilon_A + \varepsilon_D\right) \\[2mm] -\dfrac{\varepsilon_C}{2} & -\dfrac{1}{4}\left(\varepsilon_C + \varepsilon_D\right) & -\dfrac{\varepsilon_D}{2} \end{pmatrix} u = 0$$

(4.101)

5.   NODAL EQUATIONS AND MATRIX ASSEMBLY

There are 3 levels of mesh node numbering in the finite element method: local element, local nodal and global. They are illustrated in Figures 4.28 a-c.

Consider once again the six right-angle triangles shown in Figure 4.25. The residual equation is obviously localised by the shape functions of these triangles. For each triangle

$$u(x, y) = \sum_j N_j(x, y)u_j \qquad j = 1, 2, 3$$

(4.102)

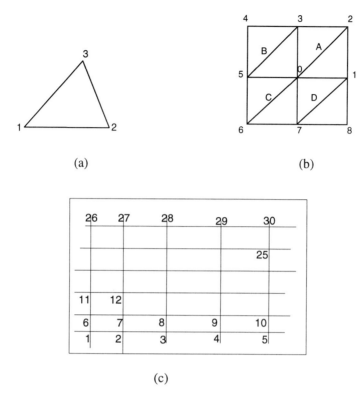

Figure 4.26 Levels of node numbering in the Finite Element Method
(a) Local (b) Nodal (c) Global

Let the global node number of the centre node 0 be node i. For each surrounding triangle, there is a contribution to the residual at node i, so

$$R_{T1} = \iint_{T1} \nabla N_i \cdot \nabla (\sum_j N_j u_j) dxdy$$

....

(4.103)

....

$$R_{T6} = \iint_{T6} \nabla N_i \cdot \nabla (\sum_j N_j u_j) dxdy \qquad j = 1,2,3$$

where R_{T1}.......R_{T6} represent the residual contributions from each triangle. We can simplify the expression for each residual. Take for instance the residual for T1

$$R_{T1} = \iint_{T1} \nabla N_i \cdot \nabla N_1 u_1 dxdy + \iint_{T1} \nabla N_i \cdot \nabla N_2 u_2 dxdy + \iint_{T1} \nabla N_i \cdot \nabla N_3 u_3 dxdy$$

(4.104)

Substituting in for the shape functions for the first-order triangle from equation (4.65)

$$R_{T1} = \sum_j K_{ij} u_j \quad \text{where} \quad K_{ij} = \frac{1}{4A_{T1}}\left(b_i b_j + c_i c_j\right) \tag{4.105}$$

The residual contributions at node i then become

$$\left[\sum_j K_{ij} u_j\right]_{T1} + \left[\sum_j K_{ij} u_j\right]_{T2} + \ldots\ldots\ldots\left[\sum_j K_{ij} u_j\right]_{T6} = 0 \quad j = 0, 1, 2, 3 \tag{4.106}$$

This is the nodal equation for the finite element method. Note that it is expressed in terms of local element numbering. In this simple case of linear function variations over triangle elements, no integration is required. In general, for higher order function variations or other element types, integration over the element area is required, but the form of this nodal equation stays the same.

For higher order variations on a triangle or other element shapes the form of the residual equation is the same but

$$K_{ij} = \iint_{A_e} \nabla N_i . \nabla N_j dxdy = \iint_{A_e} \varepsilon\left[\left(\frac{\partial N_i}{\partial x}\right)\left(\frac{\partial N_j}{\partial x}\right) + \left(\frac{\partial N_i}{\partial y}\right)\left(\frac{\partial N_j}{\partial y}\right)\right]dxdy \tag{4.107}$$

where A_e represents the area of the element. Normally numerical integration is used (see Appendix 1).

For the more general case of a derivative boundary condition on the outer surface S, $q=\partial u/\partial n$, the surface term in equation (4.92) is retained and the nodal equation is

$$\left[\sum_j K_{ij} u_j + G_i\right]_{A_{e1}} + \left[\sum_j K_{ij} u_j + G_i\right]_{A_{e2}} + \ldots\ldots\ldots\left[\sum_j K_{ij} u_j + G_i\right]_{A_{em}} = 0$$

where

$$K_{ij} = \iint_{A_e} \varepsilon\left[\left(\frac{\partial N_i}{\partial x}\right)\left(\frac{\partial N_j}{\partial x}\right) + \left(\frac{\partial N_i}{\partial y}\right)\left(\frac{\partial N_j}{\partial y}\right)\right]dxdy \tag{4.108}$$

$$G_i = -\oint \varepsilon N_i q dS$$

and m denotes the number of elements surrounding the centre node. For Poisson's equation, with a charge distribution of $-\rho(x,y)$, the term G_i is obviously modified to

$$G_i = -\iint_{A_e} N_i \rho dxdy - \oint \varepsilon N_i q dS \tag{4.109}$$

The same result can be found by the variational approach. However, in this case the starting functional W to be minimised is

$$W = \iint_{\Omega} \left(\frac{1}{2}\varepsilon |\nabla u|^2 - \rho u \right) d\Omega - \oint_{S} q u dS \qquad (4.110)$$

An interpretation of this functional is a little more complicated than before. It now consists of two terms: the stored energy in the domain plus the energy required to deposit the charge distribution in position.

After the local element to local nodal numbering translation is made, the nodal equation takes the following form,

$$\sum_k p_k u_k = 0 \qquad k = 0, 1, 2, \ldots 8 \qquad (4.111)$$

The coefficients p_k are dependent on the shapes of the elements surrounding node 0 and their material properties.

After formulating the nodal equation for each mesh node in the domain, the potential distribution can be solved by iterative methods such as successive over-relaxation, conjugate gradients, pre-conditioned conjugate gradients, or inverting a matrix by Gaussian elimination. The final matrix equation is formed by assembling the nodes according to their global node number. The form of the final matrix equation is

$$[A][u] = [B] \qquad (4.112)$$

If there are n free nodes in the domain, then matrix [A] will be of rank n by n, and matrix [B] will be of rank 1 by n. Matrix [B] is generated from the prescribed boundary conditions. Matrix A is usually termed the "stiffness" matrix. Each row of the matrix equation corresponds to a single nodal equation. A typical row of the matrix [A] takes the following form

$$(\ldots 00 \ldots P_6 P_7 P_8 \ldots 00 \ldots P_5 P_0 P_1 00 \ldots P_4 P_3 P_2 \ldots 00 \ldots)$$

where P_0 is the diagonal entry in the row.

Consider right-angle triangle elements where each triangle is assumed to have the same dielectric permittivity, as shown in Figure 4.27.

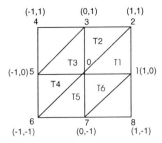

Figure 4.27 Example right-angle triangle element layout

Here the numbers in brackets represent the spatial coordinate of each node. Node 0 is assumed to lie at (0,0). It is easy to verify that for every triangle in terms of its local element node numbers

$$[K_{ij}] = \frac{\varepsilon}{2}\begin{pmatrix} 2 & -1 & -1 \\ -1 & 1 & 0 \\ -1 & 0 & 1 \end{pmatrix}$$ (4.113)

Then the nodal equation becomes

$$[K_{31} \quad K_{32} \quad K_{33}]_{T1}\begin{bmatrix} u_1 \\ u_2 \\ u_0 \end{bmatrix} + [K_{21} \quad K_{22} \quad K_{23}]_{T2}\begin{bmatrix} u_3 \\ u_0 \\ u_2 \end{bmatrix} + [K_{11} \quad K_{12} \quad K_{13}]_{T3}\begin{bmatrix} u_0 \\ u_3 \\ u_5 \end{bmatrix} +$$

$$[K_{31} \quad K_{32} \quad K_{33}]_{T4}\begin{bmatrix} u_5 \\ u_6 \\ u_0 \end{bmatrix} + [K_{21} \quad K_{22} \quad K_{23}]_{T5}\begin{bmatrix} u_7 \\ u_0 \\ u_6 \end{bmatrix} + [K_{11} \quad K_{12} \quad K_{13}]_{T6}\begin{bmatrix} u_0 \\ u_7 \\ u_1 \end{bmatrix} = 0$$ (4.114)

substituting in for the appropriate K_{ij} for each triangle gives

$$u_1 + u_3 + u_5 + u_7 - 4u_0 = 0$$ (4.115)

This result is identical to the one obtained by 5pt finite difference.

The finite element method can be summarized in the following way:

1) Generate the numerical mesh

2) Calculate the local element stiffness matrix entries, K_{ij}

3) Calculate the coefficients P_0......P_8 for the local nodal equation

4) Account for the boundary conditions into a matrix [B]

5) Form the global matrix   [A][u]=[B], where the rows of this matrix equation are derived from the nodal equations

6) Solve the matrix equation by Gaussian elimination or a preconditioned conjugate gradient method

6.   AXISYMMETRIC CYLINDRICAL COORDINATES

6.1   The general nodal equation

For systems having cylindrical symmetry in which the potential function u depends only on the radial and longitudinal coordinates (r, z), stiffness matrix entries are given by

$$K_{ij} = 2\pi \iint_{A_e} \varepsilon r \left[\left(\frac{\partial N_i}{\partial r} \right)\left(\frac{\partial N_j}{\partial r} \right) + \left(\frac{\partial N_i}{\partial z} \right)\left(\frac{\partial N_j}{\partial z} \right) \right] drdz \qquad (4.116)$$

where r the radius can also be expanded in terms of the shape functions,

$$r = \sum_{j=1}^{n} N_j r_j \qquad (4.117)$$

The K_{ij} coefficients are usually calculated by numerical integration (see Appendix 1).

6.2   First-order Elements

For linear function variations across a triangle element, let the radii at the local vertices 1, 2 and 3 be given by r_1, r_2, and r_3. The shape functions are

$$N_j(z,r) = \frac{1}{2A_e}\left(a_j + b_j z + c_j r \right) \quad j = 1, 2, 3 \qquad (4.118)$$

The coefficients a_j, b_j and c_j are found by using equation (4.65). The local stiffness matrix entries K_{ij} are given by

$$K_{ij} = \frac{\pi \varepsilon r_{av}}{2A_e}\left(b_i b_j + c_i c_j \right) \quad \text{where} \quad r_{av} = \frac{1}{3}\left(r_1 + r_2 + r_3 \right) \qquad (4.119)$$

Here the property that $N_j(r,z)$ integrated over the triangle is equal to $A_e/3$ is used.

Unlike the situation in two dimensional rectilinear coordinates, the end result depends on the way the region around the ith node is divided up into triangle elements. Consider the two orthogonal triangle element layouts A and B, as shown in Figure 4.28.

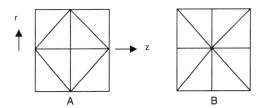

Figure 4.28 Two orthogonal triangle element layouts

It is straightforward to show that on a regularly spaced mesh, if the permittivity for each element is the same, the finite element picture stars are

$$\left(\nabla^2 u\right)_A \approx \left(\begin{array}{ccc} & -\left(1+\dfrac{h}{3r_0}\right) & \\ -1 & 4 & -1 \\ & -\left(1-\dfrac{h}{3r_0}\right) & \end{array} \right) u \qquad \left(\nabla^2 u\right)_B \approx \left(\begin{array}{ccc} & -\left(1+\dfrac{2h}{3r_0}\right) & \\ -1 & 4 & -1 \\ & -\left(1-\dfrac{2h}{3r_0}\right) & \end{array} \right) u \qquad (4.120)$$

where h is the mesh spacing and r_0 is the radius at the centre node. These two picture stars are slightly different to the equivalent 5pt finite difference expression which is given by

$$\left(\begin{array}{ccc} & 1+\dfrac{h}{2r_0} & \\ 1 & -4 & 1 \\ & 1-\dfrac{h}{2r_0} & \end{array} \right) u_{ij} \qquad (4.121)$$

However, it is simple to verify that the finite difference expression is equal to the **average** of the two finite element picture stars.

Another case of interest is the square element where the potential has variations proportional to r, z and rz. If the sides are of length h, then the following picture star can be derived

$$\frac{\pi r_0}{3} \begin{pmatrix} 1+\dfrac{h}{2r_0} & 1+\dfrac{h}{2r_0} & 1+\dfrac{h}{2r_0} \\ 1 & -8 & 1 \\ 1-\dfrac{h}{2r_0} & 1-\dfrac{h}{2r_0} & 1-\dfrac{h}{2r_0} \end{pmatrix} u \qquad (4.122)$$

This picture star is very similar to the right-angle triangle case, but differs only at the corner node positions.

6.3   On-axis nodes

Consider nodes in axisymmetric coordinates which lie on the optic axis, that is where r=0. Like the finite difference method, special attention is required since Laplace's equation in cylindrical coordinates is singular at r=0. Just as with finite difference, the following property can be used

$$\frac{1}{r}\frac{\partial u}{\partial r} \rightarrow \frac{\partial^2 u}{\partial^2 r} \qquad r \rightarrow 0 \qquad\qquad (4.123)$$

For r=0, Laplace's equation becomes

$$\nabla^2 V = 2\frac{\partial^2 u}{\partial r^2}+\frac{\partial^2 u}{\partial z^2} = 0 \qquad\qquad (4.124)$$

The weighted residual method can use this modified equation directly, alternatively, the variational method uses the modified function

$$F = \pi \iint \varepsilon \left[2\left(\frac{\partial u}{\partial r}\right)^2 + \left(\frac{\partial u}{\partial z}\right)^2 \right] r dr dz \qquad\qquad (4.125)$$

Using the modified functional given above for either of the right-angle triangle elements shown in Figure 4.28, it is simple to show that the pictorial star obtained for r=0 is identical to the one obtained by 5pt finite difference.

$$\frac{1}{h^2} \begin{pmatrix} & 4 & \\ 1 & -6 & 1 \end{pmatrix} u \qquad\qquad (4.126)$$

For the square element having potential variations proportional to r, z and rz, the on-axis picture star is

$$\frac{h\pi}{12}\begin{pmatrix} 3 & 6 & 3 \\ -1 & -10 & -1 \end{pmatrix}u \qquad (4.127)$$

Consider the more general case for two right-angle triangle elements A and B having unequal sides, as depicted in Figure 4.29.

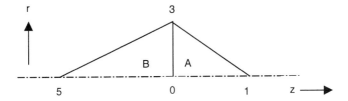

Figure 4.29 Unequal length right-angle elements on-axis

Let the distances of nodes 1, 3 and 5 from the centre node 0 be h_1, h_3 and h_5 respectively. Let A_A be the area of element A and A_B be the area of element B. Let also the dielectric permittivities of elements A and B be given by ε_A and ε_A respectively. The weighted residual or variational method taking into account the optic axis then gives the following pictorial star

$$\nabla^2 u \approx \left[\begin{array}{ccc} & -\dfrac{2}{h_3^2}\left(\Delta_A + \Delta_B\right) & \\ -\dfrac{\Delta_B}{h_5^2} & \left(\dfrac{2}{h_3^2}\left(\Delta_A + \Delta_B\right) + \dfrac{\Delta_B}{h_5^2} + \dfrac{\Delta_A}{h_1^2}\right) & -\dfrac{\Delta_A}{h_1^2} \\ & & \end{array} \right] u \qquad (4.128)$$

where $\Delta_A = A_A \varepsilon_A$ and $\Delta_B = A_B \varepsilon_B$. This picture star can then be used directly for on-axis nodes.

6.4   Near-axis r^2 correction

Let the potential $u(r,z)$ be expanded in r for a given value of z in the following way

$$u(r,z) = c_0 + c_2 r^2 + c_4 r^4 + \ldots \ldots \qquad (4.129)$$

The potential is an even function due to symmetry. Since first-order finite elements specify the potential to vary linearly, they will naturally incur inaccuracies for the radial variation of the potential close to the axis. Lencová has proposed circumventing this problem by using first-order elements in which potential variations are proportional to (r^2, z) [1]. This leads to the following energy functional for a triangle element

$$F = \frac{\pi \varepsilon_0}{6} \frac{1}{D} \sum_{i=1}^{3} \sum_{j=1}^{3} \left(\frac{3}{2} d_i d_j + 2 p c_i c_j \right) u_i u_j$$

where

$$d_1 = r_2^2 - r_3^2, \quad d_2 = r_3^2 - r_1^2, \quad d_3 = r_1^2 - r_2^2$$
$$c_1 = z_3 - z_2, \quad c_2 = z_1 - z_3, \quad c_3 = z_2 - z_1 \qquad (4.130)$$
$$p = r_1^2 + r_2^2 + r_3^2, \quad D = z_1 d_1 + z_2 d_2 + z_3 d_3$$

Minimising F with respect to u_i can then proceed in the normal way. If this method is used, there is no need to treat the nodes on axis differently to those off-axis.

A more general way of specifying the r^2 correction is to incorporate it directly into the shape function. That is, for a first-order triangle element,

$$N_j(r^2, z) = \frac{1}{\det A} \left(a_j + b_j z + c_j r^2 \right) \quad j = 1, 2, 3 \qquad (4.131)$$

where detA is the determinant of the following matrix A

$$A = \begin{pmatrix} 1 & z_1 & r_1^2 \\ 1 & z_2 & r_2^2 \\ 1 & z_3 & r_3^2 \end{pmatrix} \qquad (4.132)$$

7.   EDGE ELEMENTS

A new class of elements have recently been reported which have considerable advantages in the computation of electromagnetic wave problems. Since it is mainly static field applications of the finite element which are of interest in this book, edge elements will only be mentioned here in passing. Edge elements are reported to be a way of eliminating the troublesome "spurious" mode difficulty in computational electromagnetics. This problem arises from the lack of enforcement of the divergence condition on the field, and gives rise to the generation of many non-physical modes. Edge elements are also effective in curing the singularity problem generated by conductor or material corners [2, 3].

The concept of edge elements is based upon solving the field distribution in terms of vectors at element edges rather than field magnitudes at element nodes. Consider the rectangle element shown in Figure 4.30. Here the tangential fields E_{x1}, E_{x2}, E_{y1} and E_{y2} at the element edges are the unknown quantities to be solved. If the rectangle has lengths d_x and d_y in the x and y directions respectively, then first-order expressions for field components in the element interior are given by

$$E_x = \frac{1}{d_y}\left(y_c + \frac{d_y}{2} - y\right)E_{x1} + \frac{1}{d_y}\left(y - y_c + \frac{d_y}{2}\right)E_{x2}$$

$$E_y = \frac{1}{d_x}\left(x_c + \frac{d_x}{2} - x\right)E_{y1} + \frac{1}{d_x}\left(x - x_c + \frac{d_x}{2}\right)E_{y2}$$

(4.133)

where x_c and y_c are the coordinates for the element centre.

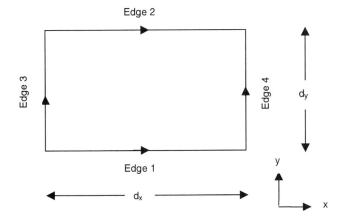

Figure 4.30 Two dimensional rectangle edge element

Like the situation for nodal elements, the unknown function distribution can be expanded in terms of basis functions, only in this case, the shape functions are vector quantities. In vector form, if **E** is the local electric field vector in the element, then obviously

$$\mathbf{E} = \sum_{j=1}^{4} \mathbf{N}_j E_j$$

(4.134)

where E_j is the tangential electric field along edge j and $\mathbf{N}_j$ are the vector shape functions given by

$$\mathbf{N}_1 = \frac{1}{d_y}\left(y_c + \frac{d_y}{2} - y\right)\hat{x} \qquad \mathbf{N}_2 = \frac{1}{d_y}\left(y - y_c + \frac{d_y}{2}\right)\hat{x}$$

$$\mathbf{N}_3 = \frac{1}{d_x}\left(x_c + \frac{d_x}{2} - x\right)\hat{y} \qquad \mathbf{N}_4 = \frac{1}{d_x}\left(x - x_c + \frac{d_x}{2}\right)\hat{y}$$

(4.135)

The assembly and formulation of the problem then proceeds in a similar way to nodal elements. Note that in general, there are more edges than nodes, so that the number of unknowns is increased. On the other hand, there is less coupling between edges, so that the final global matrix will be more sparse. In practice, the computational effort to solve an edge element problem does not significantly differ from its nodal equivalent.

The edge element for a triangle in (x,y) coordinates is shown in Figure 4.31. Here the edge element shape functions N_1, N_2 and N_3 can be conveniently expressed in terms of the first-order triangle nodal shape functions L_1, L_2, and L_3 given by equation (4.65)

$$N_1 = (L_1 \nabla L_2 - L_2 \nabla L_1) d_1$$
$$N_2 = (L_2 \nabla L_3 - L_3 \nabla L_2) d_2 \qquad\qquad (4.136)$$
$$N_3 = (L_3 \nabla L_1 - L_1 \nabla L_3) d_3$$

where d_1, d_2 and d_3 are the lengths of edges 1,2 and 3 respectively. The reader can check that $\nabla \cdot N_j = 0$ for the shape functions, and thus the divergence of the field is automatically satisfied. Other element types, three dimensional and higher-order can be derived in a similar way.

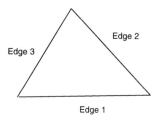

Edge 2

Edge 3

Edge 1

Figure 4.31 Triangle Edge Element

REFERENCES

[1]   B. Lencová, "Computation of electrostatic lenses and multipoles by the first-order finite element method", Nuclear Instruments and Methods in Physics Research A, 363, 1995, p190-7

[2]   P. P. Silvester and G. Pelosi, "Finte Elements for Wave Electromagnetics, methods and techniques", IEEE press, NY, 1994

[3]   Jianming Jin, "The Finite Element Method in Electromagnetics", John Wiley and Sons, Inc., New York, 1993

Chapter 5
HIGH-ORDER ELEMENTS

Elements which utilize higher than linear basis function variations of the unknown potential are well known and have been extensively studied over the last three decades. This is an instance where the use of finite elements in charged particle optics lags quite far behind its use in other subjects such as civil and mechanical engineering. Detailed studies have shown that in general, higher-order elements are more accurate than first-order elements, but that this advantage depends on the precise problem under investigation [1].

Recently, second-order elements have been proposed to be a general way of improving finite element accuracy in electron optics [2]. They have been compared to first-order elements for a few test examples and found to be more accurate. This has set off a controversy over which type of element is more accurate. Also, third-order elements to solve the axisymmetric electrostatic fields of electron guns have been used, but no detailed comparison with first-order or second-order elements was made [3]. As yet, very little reference has been made to the considerable amount of experience accumulated on this subject outside charged particle optics. Elsewhere, the transition from first-order to higher-order elements is actually automated in a hierarchical way within a single program [4]. Some others suggest using first-order elements as a means of obtaining an initial trial solution, and then switching to second-order elements to obtain greater accuracy [5].

The following section will give an introduction to the formulation of some high-order elements. They will not only be presented in general form, but in some instances, a simplified picture star will be presented. This is so that the reader can readily implement them for simple test examples and compare their accuracy to first-order elements or finite difference methods.

1. TRIANGLE ELEMENTS

The second-order triangle element, like its first-order counterpart, offers greater flexibility in modeling boundary shapes. In addition to the corner nodes, it has mid-side nodes, as illustrated in Figure 5.1.

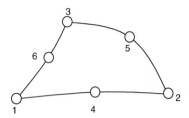

Figure 5.1 Second-order triangle element

An attractive feature about this element is that its six shape functions, $N_1,...N_6$, can be formulated in terms of the shape functions of the first-order triangle element, where the corner nodes (1,2 and 3), provide L_1, L_2, and L_3. The shape functions are given by

$$u = \sum_{j}^{6} N_j u_j \qquad x = \sum_{j}^{6} N_j x_j \qquad y = \sum_{j}^{6} N_j y_j$$

where

$$N_1 = (2L_1-1)L_1 \quad N_2 = (2L_2-1)L_2 \quad N_3 = (2L_3-1)L_3 \quad N_4 = 4L_1L_2$$
$$N_5 = 4L_2L_3 \qquad N_6 = 4L_1L_3 \tag{5.1}$$

Since L_1, L_2, and L_3 have a linear dependence on (x,y), then both the potential distribution and shape of the element boundaries will have quadratic variations in (x,y). It is convenient to take the shape functions L_1 and L_2 as normalised coordinates. Their role as local element invariants, independent of nodal coordinates helps in simplifying the calculation of K_{ij}, the entries in the element stiffness matrix. In (x,y) coordinates over the element area Ω_e

$$K_{ij} = \int_{\Omega_e} \left[\frac{\partial N_i}{\partial x} \frac{\partial N_j}{\partial x} + \frac{\partial N_i}{\partial y} \frac{\partial N_j}{\partial y} \right] d\Omega_e = \int_0^1 \int_0^{1-L_1} \left[\frac{\partial N_i}{\partial x} \frac{\partial N_j}{\partial x} + \frac{\partial N_i}{\partial y} \frac{\partial N_j}{\partial y} \right] |detJ| dL_1 dL_2$$

$$(5.2)$$

The determinant $|detJ|$ and the derivatives of the shape functions are calculated from

$$\begin{pmatrix} \dfrac{\partial N_i}{\partial x} \\[2ex] \dfrac{\partial N_i}{\partial y} \end{pmatrix} = J^{-1} \begin{pmatrix} \dfrac{\partial N_i}{\partial L_1} \\[2ex] \dfrac{\partial N_i}{\partial L_2} \end{pmatrix} \qquad J = \begin{pmatrix} \dfrac{\partial x}{\partial L_1} & \dfrac{\partial y}{\partial L_1} \\[2ex] \dfrac{\partial x}{\partial L_2} & \dfrac{\partial y}{\partial L_2} \end{pmatrix} \qquad (5.3)$$

These expressions can be integrated analytically or numerically. Appendix 1 gives a useful integration formula for triangles when analytical integration is sought. The foregoing approach is perfectly general and is independent of the element's nodal coordinates. It is instructive to examine the pictorial star for the second-order triangle element on a regularly spaced mesh. Consider the right-angled triangle element, as shown in Figure 5.2 and let the area around a node be divided up in two orthogonal ways, as indicated by A and B in Figure 5.3.

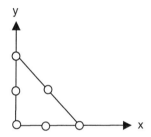

Figure 5.2 Right angle triangle

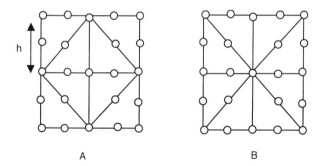

Figure 5.3 Two orthogonal triangle element layouts

Layouts A and B can of course be combined together. The sides of each triangle element in the x or y directions have equal length. Assuming that the nodes between the corners lie at the middle of a side, then it is straightforward to show that for

Laplace's equation in the potential u(x,y), the pictorial star at the centre node of either layout A or B is given by

$$\left(\begin{array}{ccccc} & & \dfrac{1}{12} & & \\ & & -\dfrac{1}{3} & & \\ \dfrac{1}{12} & -\dfrac{1}{3} & 1 & -\dfrac{1}{3} & \dfrac{1}{12} \\ & & -\dfrac{1}{3} & & \\ & & \dfrac{1}{12} & & \end{array}\right) u \tag{5.4}$$

This picture star can be viewed as a weighted average of two first-order finite element or 5pt finite difference picture stars which correspond to the Richardson's extrapolation scheme

$$\frac{4}{3}\left(\begin{array}{ccc} & -1 & \\ -1 & 4 & -1 \\ & -1 & \end{array}\right) u - \frac{1}{3}\left(\begin{array}{ccccc} & & -1 & & \\ & & 0 & & \\ -1 & 0 & 4 & 0 & -1 \\ & & 0 & & \\ & & -1 & & \end{array}\right) u \tag{5.5}$$

There is however an important difference. The second-order triangle element picture star corresponds to a row of the global matrix equation before a solution is found, while Richardson's extrapolation is applied to two potential distributions which are already solved, that is, it is a post-processing operation. For mid-side nodes, the picture star turns out to be identical to the standard 5pt difference star, that is, one where the nearest neighboring nodes in the x and y direction are multiplied by $-1/4$.

In axi-symmetric coordinates (r,z), an additional r term complicates the calculation of the relevant K_{ij} terms. The radius is expressed in terms of the shape functions, so

$$K_{ij} = \int_{\Omega_e} \left[\frac{\partial N_i}{\partial r}\frac{\partial N_j}{\partial r} + \frac{\partial N_i}{\partial z}\frac{\partial N_j}{\partial z}\right] d\Omega_e = 2\pi \int_0^1 \int_0^{1-L_1} \left[\frac{\partial N_i}{\partial r}\frac{\partial N_j}{\partial r} + \frac{\partial N_i}{\partial z}\frac{\partial N_j}{\partial z}\right] r |\det J| dL_1 dL_2$$

where

$$r = \sum_{j=1}^{6} N_j r_j \tag{5.6}$$

Since numerical integration is usually used, very little extra effort is required to calculate K_{ij} in the axi-symmetric case. Just as with first-order triangle elements, the result is dependent on the way the surrounding space is divided into elements. Consider the two orthogonal mesh layouts A and B depicted in Figure 5.3. Here all elements are of the same size, right-angled, and have a side length of h. Let R be the radius of the centre node. On layout A, the pictorial star for the centre node is given by

$$
\begin{pmatrix}
 & & \dfrac{h}{30R}+\dfrac{1}{12} & & \\[2mm]
 & \dfrac{h}{40R} & -\dfrac{1.4h}{12R}-\dfrac{1}{3} & \dfrac{h}{40R} & \\[2mm]
\dfrac{1}{12} & -\dfrac{1}{3} & 1 & -\dfrac{1}{3} & \dfrac{1}{12} \\[2mm]
 & -\dfrac{h}{40R} & \dfrac{1.4h}{12R}-\dfrac{1}{3} & -\dfrac{h}{40R} & \\[2mm]
 & & -\dfrac{h}{30R}+\dfrac{1}{12} & &
\end{pmatrix} u
\tag{5.7}
$$

On the other hand, the pictorial star for layout B involves 8 more nodes

$$
\begin{pmatrix}
 & \dfrac{h}{40R} & \dfrac{h}{20R}+\dfrac{1}{12} & \dfrac{h}{40R} & \\[2mm]
\dfrac{h}{60R} & -\dfrac{h}{24R} & -\dfrac{1.8h}{12R}-\dfrac{1}{3} & -\dfrac{h}{24R} & \dfrac{h}{60R} \\[2mm]
\dfrac{1}{12} & -\dfrac{1}{3} & 1 & -\dfrac{1}{3} & \dfrac{1}{12} \\[2mm]
-\dfrac{h}{60R} & \dfrac{h}{24R} & \dfrac{1.8h}{12R}-\dfrac{1}{3} & \dfrac{h}{24R} & -\dfrac{h}{60R} \\[2mm]
 & -\dfrac{h}{40R} & -\dfrac{h}{20R}+\dfrac{1}{12} & -\dfrac{h}{40R} &
\end{pmatrix} u
\tag{5.8}
$$

As with first-order finite elements the two layouts can added together and averaged. The result is expected to be more accurate. The averaged pictorial star becomes

$$
\begin{pmatrix}
 & \dfrac{h}{80R} & \dfrac{h}{24R}+\dfrac{1}{12} & \dfrac{h}{80R} & \\[2mm]
\dfrac{h}{120R} & -\dfrac{h}{120R} & -\dfrac{2h}{15R}-\dfrac{1}{3} & -\dfrac{h}{120R} & \dfrac{h}{120R} \\[2mm]
\dfrac{1}{12} & -\dfrac{1}{3} & 1 & -\dfrac{1}{3} & \dfrac{1}{12} \\[2mm]
-\dfrac{h}{120R} & \dfrac{h}{120R} & \dfrac{2h}{15R}-\dfrac{1}{3} & \dfrac{h}{120R} & -\dfrac{h}{120R} \\[2mm]
 & -\dfrac{h}{80R} & -\dfrac{h}{24R}+\dfrac{1}{12} & -\dfrac{h}{80R} &
\end{pmatrix} u
\tag{5.9}
$$

Note that as R increases in magnitude, this pictorial star becomes identical to the previous one in (x,y) coordinates. Note also that the nearest diagonal terms to the centre node have been substantially reduced in magnitude by forming the average between the two layouts. This is quite similar to the corresponding situation for first-order triangle elements where the average between orthogonal mesh layouts had the effect of eliminating the corner terms. All cross-terms in the finite difference case do not appear, while for the second-order averaged triangle element, they are relatively small. Similarities are also apparent for the pictorial stars relating to the mid-side nodes. Take for instance the mid-side node 0 in the z-direction of the 8 second-order triangle elements shown in Figure 5.4.

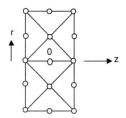

Figure 5.4 Mid-side node of 8 triangles

The picture star in this case, an average of two sets of four second-order triangles, is given by

$$
\left(
\begin{array}{ccc}
-\dfrac{3h}{320\ R} & & -\dfrac{3h}{320\ R} \\[2mm]
\dfrac{h}{80\ R} & -\dfrac{1}{4}\left(1+\dfrac{h}{5R}\right) & \dfrac{h}{80\ R} \\[2mm]
-\dfrac{1}{4} & 1 & -\dfrac{1}{4} \\[2mm]
-\dfrac{h}{80\ R} & -\dfrac{1}{4}\left(1-\dfrac{h}{5R}\right) & -\dfrac{h}{80\ R} \\[2mm]
\dfrac{3h}{320\ R} & & \dfrac{3h}{320\ R}
\end{array}
\right) u
\tag{5.10}
$$

This pictorial star is similar to the corresponding 5 pt finite difference star centred at node 0 with mesh spacing h/2

$$
\left(
\begin{array}{ccc}
 & -\dfrac{1}{4}\left(1+\dfrac{h}{4R}\right) & \\[2mm]
-\dfrac{1}{4} & 1 & -\dfrac{1}{4} \\[2mm]
 & -\dfrac{1}{4}\left(1-\dfrac{h}{4R}\right) &
\end{array}
\right) u
\tag{5.11}
$$

Similarities also exist for the other mid-side nodes. Note that the nodal equation for the mid-side nodes are not in this case all equal. The nodal equation depends on whether the mid-side node lies along the r or z direction.

The nodal scheme for the cubic triangle element is illustrated in Figure 5.5 below.

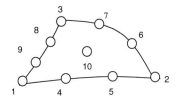

Figure 5.5 Cubic triangle element

The shape functions have values of 1/3 and 2/3 at the edge nodes. For the corner nodes are

$$N_1 = \frac{1}{2}(3L_1 - 1)(3L_1 - 2)L_1 \quad \text{Etc.}$$

For mid-side nodes

$$N_4 = \frac{9}{2}(3L_1 - 1)L_1L_2 \quad N_5 = \frac{9}{2}(3L_1 - 2)L_1L_2 \quad \text{Etc.}$$

For the internal node

$$N_{10} = 27L_1L_2L_3 \tag{5.12}$$

2. QUADRILATERAL ELEMENTS

There are several possible ways to form second-order quadrilateral shaped elements. One possible scheme is to use 4 nodes lying on element corners, 4 along the edges, and an internal node, giving a total of 9 nodes. Figure 5.6 shows the nodal layout for this element.

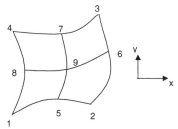

Figure 5.6 9 node second-order quadrilateral element

The shape functions in this case are given by

$$N_1(u, v) = \frac{1}{4}(1 - u)(1 - v)uv \qquad N_2(u, v) = \frac{1}{4}(1 + u)(1 - v)uv$$

$$N_3(u, v) = \frac{1}{4}(1 + u)(1 + v)uv \qquad N_4(u, v) = \frac{1}{4}(1 - u)(1 + v)uv$$

$$N_5(u, v) = \frac{1}{2}(1 - u^2)(1 - v)v \qquad N_6(u, v) = \frac{1}{2}(1 - v^2)(1 + u)u \qquad (5.13)$$

$$N_7(u, v) = \frac{1}{2}(1 - u^2)(1 + v)v \qquad N_8(u, v) = \frac{1}{2}(1 - v^2)(1 - u)u$$

$$N_9(u, v) = (1 - v^2)(1 - u^2)$$

The same procedure as before is used to calculate the element's K_{ij} terms. The shape functions and the coordinates of each node are used to calculate shape function derivatives. From a programming perspective, numerical integration is by far the simpler approach. Note that from merely topological considerations, the 9 node rectangle element can be sub-divided into four first-order rectangle elements, so that it can be used on the same kind of mesh which is generated for first-order elements. There is however, an important qualification to this observation: conductor or material boundaries must cover the entire element and not just a part of it. If they happen to partially fill the element, significant errors will arise, making the results more inaccurate than if first-order elements had been used on the same mesh. The conductor, source and material regions must therefore span an even number of mesh cells. This places a restriction on the type of mesh that can be used with the 9 node second-order element [6]. The pictorial stars for regularly spaced mesh lines can be derived analytically, and are given in Appendix 2.

3.   THE SERENDIPITY FAMILY OF ELEMENTS

There is another class of elements, and they have been named the "Serendipity family" of elements by Zienkiewicz. Here the nodes lie either on the element corner or edge. The second-order quadrilateral element is shown in Figure 5.7.

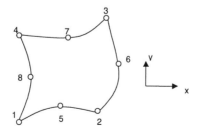

Figure 5.7 The 8 node quadrilateral element

The shape functions for this element are given by

$$N_1(u,v) = \frac{1}{4}(1-u)(1-v)(-u-v-1) \qquad N_2(u,v) = \frac{1}{4}(1+u)(1-v)(u-v-1)$$

$$N_3(u,v) = \frac{1}{4}(1+u)(1+v)(u+v-1) \qquad N_4(u,v) = \frac{1}{4}(1-u)(1+v)(-u+v-1) \quad (5.14)$$

$$N_5(u,v) = \frac{1}{2}(1-u^2)(1-v) \qquad\qquad N_6(u,v) = \frac{1}{2}(1-v^2)(1+u)$$

$$N_7(u,v) = \frac{1}{2}(1-u^2)(1+v) \qquad\qquad N_8(u,v) = \frac{1}{2}(1-v^2)(1-u)$$

The normalised coordinates u and v vary between -1 and $+1$. The corner nodes lie at either -1 or 1, while either u or v is zero for mid-side nodes.   An important characteristic of this type of element is that its shape functions can be systematically derived in terms of lower-order elements. First, the edge shape functions are written down by inspection. For example, node 5 on the second-order element is given by

$$N_5(u,v) = \frac{1}{2}\left(1-u^2\right)\left(1-v\right) \qquad\qquad (5.15)$$

Clearly this shape function is quadratic in u and linear in v. Similarly for node 8

$$N_8(u,v) = \frac{1}{2}\left(1-u\right)\left(1-v^2\right) \qquad\qquad (5.16)$$

The fact that this shape function is quadratic in v and linear in u is also apparent from inspection. For corner nodes, the shape function can be derived by using the first-order shape function, and subtracting suitable amounts of the edge shape functions, so as to make it zero along all edge nodes. Take for instance corner 1, the shape function from the first-order element $M_1(u,v)$ is given by

$$M_1(u,v) = \frac{1}{4}\left(1-u\right)\left(1-v\right) \qquad\qquad (5.17)$$

The value of this function is ½ at node 5 and 8, so $N_1(u,v)$ is derived by subtracting these amounts of nodes 5 and 8 away from $M_1(u,v)$

$$N_1(u,v) = M_1(u,v) - \frac{1}{2}N_5 - \frac{1}{2}N_8 \qquad\qquad (5.18)$$

The reader can confirm that this expression is identical to the one already presented for $N_1(u,v)$. By allowing for systematic inclusion of edge nodes, it is clear that mixed order elements can be created. Consider the situation where an extra mid-side node is added to the bottom edge of a first-order element as shown in Figure 5.8.

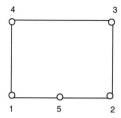

Figure 5.8 Mixed-order quadrilateral element

Using the above method, it is clear that the shape functions for such an element are given by

$$N_5(u,v) = \frac{1}{2}(1-u^2)(1-v) \quad N_3(u,v) = \frac{1}{4}(1+u)(1+v) \quad N_4(u,v) = \frac{1}{4}(1-u)(1+v)$$

$$N_1(u,v) = \frac{1}{4}(1-u)(1-v) - \frac{1}{2}N_5(u,v) \quad N_2(u,v) = \frac{1}{4}(1+u)(1-v) - \frac{1}{2}N_5(u,v)$$

$$(5.19)$$

The ability to specify mixed-order elements is a useful feature of Serendipity elements. The order can obviously be increased according to a local error estimator, allowing for element order refinement. For electron lenses, an obvious region where higher-order elements may be required is close to the optic axis. This particularly useful for first-order elements in axisymmetric coordinates (r,z) which requires the potential to vary with r^2 instead of linearly with r in the near-axis region. By using second-order elements for the first few mesh cells in the radial direction, the r^2 requirement is automatically satisfied.

The cubic element has two edge nodes, in terms of the normalised coordinates (u,v), they are located at −1/3 and 1/3 along the edge direction. The nodal layout for the element is illustrated in Figure 5.9.

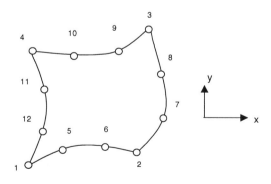

Figure 5.9 A 12 node cubic quadrilateral element

The shape functions along the edges are simple to derive by inspection. Take for instance node 5 located at u=-1/3 and v=-1. It must be cubic in u and linear v. To be zero at the corners 1 and 2, the term $(1-u^2)$ must be used. To be zero at node 6 (u=1/3) requires the inclusion of the term $(1-3u)$. So the shape function at node 5 is given by

$$N_5 = \frac{9}{32}(1-v)(1-u^2)(1-3u)$$

(5.20)

Likewise for the other edge nodes

$$N_6 = \frac{9}{32}(1-v)(1-u^2)(1+3u) \qquad N_9 = \frac{9}{32}(1+v)(1-u^2)(1+3u)$$

$$N_{10} = \frac{9}{32}(1+v)(1-u^2)(1-3u) \qquad N_7 = \frac{9}{32}(1+u)(1-v^2)(1-3v)$$

(5.21)

$$N_8 = \frac{9}{32}(1+u)(1-v^2)(1+3v) \qquad N_{11} = \frac{9}{32}(1-u)(1-v^2)(1+3v)$$

$$N_{12} = \frac{9}{32}(1-u)(1-v^2)(1-3v)$$

The shape functions at the corners can be derived either in terms second-order elements or by inspection. They are as follows

$$N_1(u,v) = \frac{1}{32}(1-u)(1-v)[-10+9(u^2+v^2)]$$

$$N_2(u,v) = \frac{1}{32}(1+u)(1-v)[-10+9(u^2+v^2)]$$

(5.22)

$$N_3(u,v) = \frac{1}{32}(1+u)(1+v)[-10+9(u^2+v^2)]$$

$$N_4(u,v) = \frac{1}{32}(1-u)(1+v)[-10+9(u^2+v^2)]$$

The principle of mixing element order presented earlier also applies to deriving the shape functions for high-order mixed elements. Consider the first-order/cubic element shown in Figure 5.10.

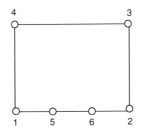

Figure 5.10 First-order/cubic mixed-order element

Here the corner shape function for nodes 1 and 2 are given by

$$N_1(u,v) = M_1 - \frac{2}{3}N_5 - \frac{1}{3}N_6$$

$$N_2(u,v) = M_2 - \frac{1}{3}N_5 - \frac{2}{3}N_6$$

(5.23)

where M_1 and M_2 are the first-order shape functions for nodes 1 and 2 respectively, and N_5 and N_6 are cubic shape functions.

REFERENCES

[1]   P. P. Silvester and R. L. Ferrari, "Finite Elements for Electrical Engineers", Cambridge University Press, England, 2nd ed., 1983, Chapter 3, section 7, p88-90

[2]   X. Zhu and E. Munro, "Second-order finite element method and its practical application in charged particle optics", Journal of Microscopy, vol. 170, Pt. 2, August 1995, pp170-180

[3]   K. Saito and Y. Uno, "Accurate electron ray tracing for analysis of electron guns immersed in a magnetic lens field", Nuclear Instruments and Methods in Physics Research A, 363, 1995, pp48-53

[4]   O. C. Zienkiewicz and A. Craig, ,"Adaptive refinement, error estimates, multigrid solution, and hierarchic finite element concepts", chapter 2 in "Accuracy estimates and adaptive refinements in finite element computations", edited by Babuska, O. C. Zienkiewicz, J. Gago, E. R. de A. Oliveira,John Wiley and Sons, 1986, NY, pp25-59

[5]   K. J. Binns, P. J. Lawrenson and C. W. Trowbridge, "The analytical and numerical solution of electric and magnetic fields", John Wiley and Sons, 1992, NY, p329

[6] O. C. Zienkiewicz, "The finite element method", Mcgraw-Hill company, 3rd edition, 1977, London, p155-160

Chapter 6
ELEMENTS IN THREE DIMENSIONS

In three dimensions, the finite element formulation of electrostatic problems is similar to that used in two dimensions. Consider the Cartesian coordinate system (x,y,z), then the nodal equation for the m elements around the ith node is given by

$$\left[\sum_j K_{ij}u_j\right]_{\Omega_{e_1}} + \left[\sum_j K_{ij}u_j\right]_{\Omega_{e_2}} + \ldots\ldots\ldots\ldots\left[\sum_j K_{ij}u_j\right]_{\Omega_{e_m}} = 0$$

(6.1)

where

$$K_{ij} = \iint_{\Omega_e}\varepsilon\left[\left(\frac{\partial N_i}{\partial x}\right)\left(\frac{\partial N_j}{\partial x}\right)+\left(\frac{\partial N_i}{\partial y}\right)\left(\frac{\partial N_j}{\partial y}\right)+\left(\frac{\partial N_i}{\partial z}\right)\left(\frac{\partial N_j}{\partial z}\right)\right]dxdydz$$

where $\Omega_1\ldots\ldots\Omega_m$ are the respective volumes for each element.

On a structured mesh, one where mesh lines are defined in the x,y and z directions, the general nodal layout has 26 nodes which surround the centre node, as shown in Figure 6.1. Each plane in the z-direction has 9 nodes.

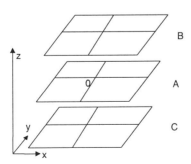

Figure 6.1 Structured nodal arrangement in three dimensions

The schematic representation shown in Figure 6.1 does not mean that the x-y planes, A, B and C are flat, it only indicates the general topology of the surrounding nodes. Each row of the global matrix is of the following form

$$\sum_{j=0}^{26} P_{ij} u_j = b_i \qquad (6.2)$$

where i denotes the ith node (global numbering), and b$_i$ is a source term which is generated either by the boundary conditions or from the presence of a fixed charge distribution. Where b$_i$ comes from a fixed potential boundary condition, the corresponding P$_{ij}$ on the left-side must of course be assigned to a zero value. As in the case for two dimensions, the global matrix is sparse and symmetric.

1.   ELEMENT SHAPE FUNCTIONS

1.1   The Brick Element

The brick element is a simple extension of the quadrilateral element. In three dimensions, it is the simplest way of dividing up the whole domain. It has eight nodes, as shown in the schematic diagram of Figure 6.2.

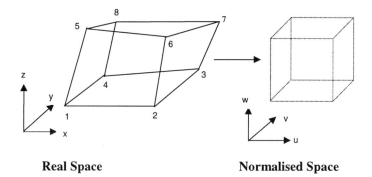

Real Space   **Normalised Space**

Figure 6.2 The Brick Element

The shape functions are given by

$$N_j(u, v, w) = \frac{1}{8}\left(1 + u_j u\right)\left(1 + v_j v\right)\left(1 + w_j w\right)$$

where (u$_j$, v$_j$ w$_j$) for j=1 to 8 is given by

1: (-1,-1,-1)   2: (1,-1,-1)   3: (1,1,-1)
4: (-1,1,-1)   5: (-1,-1,1)   6: (1,-1,1)
7: (1,1,1)   8: (-1,1,1)   (6.3)

1.2   The Tetrahedral element

One of the most popular elements in three dimensions is the tetrahedral element. It has 4 nodes and is formulated in direct analogy to the first-order triangle element. A schematic diagram of the tetrahedral element is given in Figure 6.3.

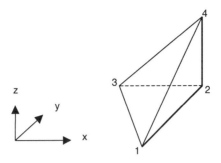

Figure 6.3 Tetrahedral element

Let the shape functions be given by $L_1(x,y,z)$, $L_2(x,y,z)$, $L_3(x,y,z)$ and $L_4(x,y,z)$. They are found by solving a matrix equation involving the (x,y,z) coordinates of the 4 nodes, $(x_1,y_1,z_1)\ldots\ldots (x_4,y_4,z_4)$

$$\begin{pmatrix} x \\ y \\ z \\ 1 \end{pmatrix} = \begin{pmatrix} x_1 & x_2 & x_3 & x_4 \\ y_1 & y_2 & y_3 & y_4 \\ z_1 & z_2 & z_3 & z_4 \\ 1 & 1 & 1 & 1 \end{pmatrix} \begin{pmatrix} L_1 \\ L_2 \\ L_3 \\ L_4 \end{pmatrix} \tag{6.4}$$

These shape functions linearly depend on x,y and z. Tetrahedral elements can of course model a wider variety of boundary shapes than brick elements, but they are more difficult to generate.

1.3   Prism elements

Prism elements are not as flexible as tetrahedral elements, but are simpler to generate. They are suitable for structures where the electrodes and dielectric regions can be modelled by straight lines in one of the x,y,z directions. Let this direction be along the z-coordinate. The prism element has 6 nodes and is shown in Figure 6.4.

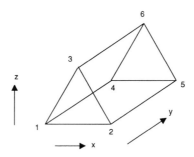

Figure 6.4 Prism Element

Since the prism element in the x-y plane consists of triangle elements which are constant with respect to the z-direction, then the shape functions, $N_1(x,y,z)$ to $N_6(x,y,z)$ can be formulated in terms of the first-order triangle shape functions $L_1(x,y)$, $L_2(x,y)$, and $L_3(x,y)$ for nodes 1, 2 and 3.

$$N_1(x,y,z) = L_1\left(1 - \frac{z}{d}\right) \quad N_2(x,y,z) = L_2\left(1 - \frac{z}{d}\right) \quad N_3(x,y,z) = L_3\left(1 - \frac{z}{d}\right)$$

$$N_4(x,y,z) = L_1\left(\frac{z}{d}\right) \quad N_5(x,y,z) = L_2\left(\frac{z}{d}\right) \quad N_6(x,y,z) = L_3\left(\frac{z}{d}\right)$$

(6.5)

where d is the distance in the z-direction separating nodes 1, 2 and 3 from the nodes 4, 5 and 6. The nodal equations can then be assembled in a manner similar to the two-dimensional case.

2. GENERATING TETRAHEDRAL ELEMENTS TO FIT CURVED BOUNDARY SURFACES

Although tetrahedral elements are used frequently in 3-D finite element analysis, the division of a region into individual tetrahedral elements that fit all boundary surfaces is a non-trivial problem, particularly when these surfaces are curved. One approach is to divide the whole domain into a structured array of eight-cornered elements, and then further subdivide each brick element into tetrahedral elements. It will be assumed here that a mesh can be generated so that the nodes of brick elements either fall on boundary surfaces, or are free nodes. A mesh generator of this type will be described later in Chapter 11, section 2. The boundary conditions are formulated in terms of how boundary surfaces intersect brick elements. Brick elements are subdivided into tetrahedral elements that fit the boundary surface. The boundary surface is approximated to be a plane within the brick element. For each brick element that has nodes on a boundary surface, tetrahedral element subdivision can be made automatically, based upon which of the brick element nodes lie on the boundary surface.

Assume that the dielectric permittivity at one side of the boundary is ε_1 and the other side is ε_2. Assume also that the dielectric permittivity for any tetrahedral element can easily be determined by using the coordinates of its vertices. Since there are several ways in which the interface plane can intersect the brick. Tetrahedral subdivision must proceed by first identifying how the interface place cuts the brick. Consider the case when the interface cuts the brick at nodes 2, 6 and 8, as shown in Figure 6.5. In this case, the brick element can be divided into five tetrahedra, T1-T5.

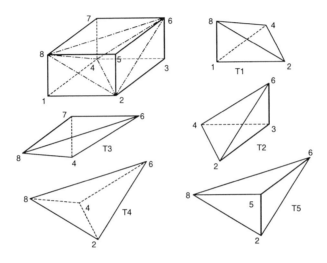

Figure 6.5 Interface plane cutting brick element at nodes 2, 6, and 8

Notice that tetrahedral T5 lies at one side of the interface, whereas the other four tetrahedral lie on the other side of the interface. This information is used to help identify the permittivity value of each tetrahedral element.

When the interface cuts the brick at nodes 1, 5 and 7, the previous subdivision generates tetrahedra which straddle the interface, hence the tetrahedral arrangement in the brick needs to be modified to the form shown in Figure 6.6.

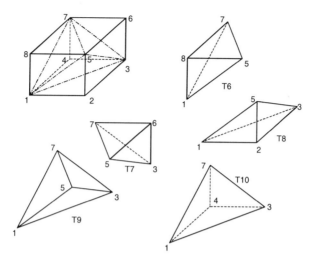

Figure 6.6 Interface plane cutting brick element at nodes 1, 5, and 7

The interface can also slice the brick at the top diagonally, splitting the brick into two prisms. Each prism can then be subdivided into three tetrahedra. There are two ways in which each prism can be divided, but this is not important since they both fit the interface plane. Consider the case when the interface cuts the brick at nodes 2, 4, 5 and 7 as shown in Figure 6.7.

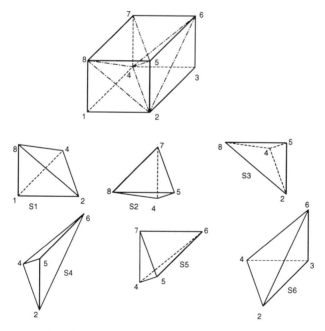

Figure 6.7 Interface plane cutting brick element at nodes 2, 4, 5, and 7

In this case S1 – S3 lies on one side of the interface, while S4 – S6 lies on the other side. It is clear that the interface plane my cut along diagonal lines at the brick's other faces, creating prisms in other directions, generating more tetrahedral subdivisions. All these possibilities can be tabulated, and in this way, depending on how the interface plane cuts the brick element, a suitable set of tetrahedral elements can always be found. Table 6.1 presents tetrahedra T1-10, S1-12 and D1-24 which account for the different ways that the interface plane can cut the brick element. Each tetrahedral is described in terms of local brick node numbers.

Tetrahedral	Local Node Numbers	Tetrahedral	Local Node Numbers
T1, S1	1,2,4,8	S9, D11	1,3,5,6
T2, S6	2,3,4,6	S11	1,3,6,7
T3, D7	4,6,7,8	S12, D22	1,6,7,8
T4	2,4,6,8	D1, D16	3,4,7,8
T5	2,5,6,8	D2, D13	1,3,4,8
T6	1,5,7,8	D3, D14	1,2,3,8
T7	3,5,6,7	D4, D17	3,6,7,8
T8, S8, D12	1,2,3,5	D5, D18	3,5,6,8
T9	1,3,5,7	D6, D15	2,3,5,8
T10, S10	1,3,4,7	D8	1,4,6,8
S2	4,5,7,8	D10, D20	1,3,4,6
S3	2,4,5,8	D19	1,4,6,7
S4	2,4,5,6	D21	1,2,3,6
S5	4,5,6,7	D24	1,2,5,6
S7, D9, D23	1,5,6,8.		

Table 6.1 Tetrahedra T1-10, S1-12, D1-24 generated by different ways an interface can cut a brick element

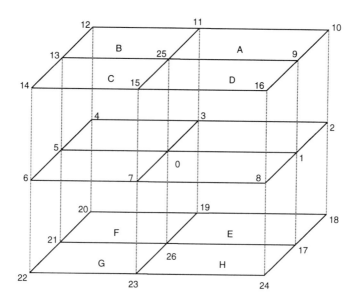

Figure 6.8 8 Brick element surrounding the centre node

Figure 6.8 illustrates how eight brick elements, labeled A-H, surround a centre node 0 on a structured three dimensional mesh. Obviously the nodal equation at 0 must take into account tetrahedra from all eight brick elements. For example, in element A, one of the tetrahedra contributing to node 0 is T1. We can then write the contribution from this tetrahedral as

$$[K_{i1}]_{A,T1} V_0 + [K_{i2}]_{A,T1} V_1 + [K_{i3}]_{A,T1} V_3 + [K_{i4}]_{A,T1} V_{25}$$

and taking the tetrahedral T2 from element B gives

$$[K_{i1}]_{B,T2} V_0 + [K_{i2}]_{B,T2} V_3 + [K_{i3}]_{B,T2} V_4 + [K_{i4}]_{B,T2} V_{11}$$

assembling the above two equations, we have

$$P_0 = [K_{i1}]_{A,T1} + [K_{i1}]_{B,T2}$$
$$P_1 = [K_{i2}]_{A,T1}$$
$$P_3 = [K_{i3}]_{A,T1} + [K_{i2}]_{B,T2}$$
$$P_4 = [K_{i3}]_{B,T2}$$
$$P_{11} = [K_{i4}]_{B,T2}$$
$$P_{25} = [K_{i4}]_{A,T1}$$

by assembling all the tetrahedral from all eight brick elements, we have

$$P_0 = \left[K_{i1}\right]_{A,T1} + \left[K_{i1}\right]_{A,T10} + \left[K_{i1}\right]_{A,S1} + \left[K_{i1}\right]_{A,S7} + \left[K_{i1}\right]_{A,S10} + \left[K_{i1}\right]_{A,S11} + \left[K_{i1}\right]_{A,S12}$$

$$+ \left[K_{i1}\right]_{A,D2} + \left[K_{i1}\right]_{A,D3} + \left[K_{i1}\right]_{A,D8} + \left[K_{i1}\right]_{A,D9} + \left[K_{i1}\right]_{A,D13} + \left[K_{i1}\right]_{A,D14} + \left[K_{i1}\right]_{A,D19}$$

$$+ \left[K_{i1}\right]_{A,D21} + \left[K_{i1}\right]_{A,D22} + \left[K_{i1}\right]_{A,D23} + \left[K_{i1}\right]_{A,D24} + \left[K_{i2}\right]_{A,T6} + \left[K_{i3}\right]_{A,T8} + \left[K_{i3}\right]_{A,S8}$$

$$+ \left[K_{i3}\right]_{A,D12} + \left[K_{i4}\right]_{A,T9} + \left[K_{i4}\right]_{A,S9} + \left[K_{i4}\right]_{A,D10} + \left[K_{i4}\right]_{A,D11} + \left[K_{i4}\right]_{A,D20} + \left[K_{i1}\right]_{B,T2}$$

$$+ \left[K_{i1}\right]_{B,T8} + \left[K_{i1}\right]_{B,S3} + \left[K_{i1}\right]_{B,S4} + \left[K_{i1}\right]_{B,S6} + \left[K_{i1}\right]_{B,S8} + \left[K_{i1}\right]_{B,D6} + \left[K_{i1}\right]_{B,D12}$$

$$+ \left[K_{i1}\right]_{B,D15} + \left[K_{i2}\right]_{B,T1} + \left[K_{i2}\right]_{B,T4} + \left[K_{i2}\right]_{B,T5} + \left[K_{i2}\right]_{B,S1} + \left[K_{i2}\right]_{B,D3} + \left[K_{i2}\right]_{B,D14}$$

$$+ \left[K_{i2}\right]_{B,D24} + \left[K_{i3}\right]_{B,D21} + \left[K_{i1}\right]_{C,D5} + \left[K_{i1}\right]_{C,D18} + \left[K_{i2}\right]_{C,T9} + \left[K_{i2}\right]_{C,S9} + \left[K_{i2}\right]_{C,S11}$$

$$+ \left[K_{i2}\right]_{C,D1} + \left[K_{i2}\right]_{C,D2} + \left[K_{i2}\right]_{C,D4} + \left[K_{i2}\right]_{C,D6} + \left[K_{i2}\right]_{C,D11} + \left[K_{i2}\right]_{C,D13} + \left[K_{i2}\right]_{C,D15}$$

$$+ \left[K_{i2}\right]_{C,D16} + \left[K_{i2}\right]_{C,D17} + \left[K_{i2}\right]_{C,D21} + \left[K_{i3}\right]_{C,T2} + \left[K_{i3}\right]_{C,T7} + \left[K_{i3}\right]_{C,T10} + \left[K_{i3}\right]_{C,S6}$$

$$+ \left[K_{i3}\right]_{C,S10} + \left[K_{i3}\right]_{C,D10} + \left[K_{i3}\right]_{C,D20} + \left[K_{i4}\right]_{C,T8} + \left[K_{i4}\right]_{C,S8} + \left[K_{i4}\right]_{C,D3} + \left[K_{i4}\right]_{C,D12}$$

$$+ \left[K_{i4}\right]_{C,D14} + \left[K_{i1}\right]_{D,T4} + \left[K_{i1}\right]_{D,S5} + \left[K_{i1}\right]_{D,D1} + \left[K_{i1}\right]_{D,D10} + \left[K_{i1}\right]_{D,D16} + \left[K_{i1}\right]_{D,D20}$$

$$+ \left[K_{i2}\right]_{D,T2} + \left[K_{i2}\right]_{D,T10} + \left[K_{i2}\right]_{D,S2} + \left[K_{i2}\right]_{D,S6} + \left[K_{i2}\right]_{D,S10} + \left[K_{i3}\right]_{D,D19} + \left[K_{i3}\right]_{D,S4}$$

$$+ \left[K_{i3}\right]_{D,D8} + \left[K_{i4}\right]_{D,T1} + \left[K_{i4}\right]_{D,T3} + \left[K_{i4}\right]_{D,S1} + \left[K_{i4}\right]_{D,S3} + \left[K_{i4}\right]_{D,D2} + \left[K_{i4}\right]_{D,D7}$$

$$+ \left[K_{i4}\right]_{D,D13} + \left[K_{i1}\right]_{E,T6} + \left[K_{i1}\right]_{E,S2} + \left[K_{i1}\right]_{E,D4} + \left[K_{i1}\right]_{E,D17} + \left[K_{i2}\right]_{E,T3} + \left[K_{i2}\right]_{E,D7}$$

$$+ \left[K_{i2}\right]_{E,D8} + \left[K_{i3}\right]_{E,T1} + \left[K_{i3}\right]_{E,T4} + \left[K_{i3}\right]_{E,S1} + \left[K_{i3}\right]_{E,S3} + \left[K_{i3}\right]_{E,S7} + \left[K_{i3}\right]_{E,S12}$$

$$+ \left[K_{i3}\right]_{E,D1} + \left[K_{i3}\right]_{E,D2} + \left[K_{i3}\right]_{E,D3} + \left[K_{i3}\right]_{E,D9} + \left[K_{i3}\right]_{E,D13} + \left[K_{i3}\right]_{E,D14} + \left[K_{i3}\right]_{E,D16}$$

$$+ \left[K_{i3}\right]_{E,D22} + \left[K_{i3}\right]_{E,D23} + \left[K_{i4}\right]_{E,T5} + \left[K_{i4}\right]_{E,D5} + \left[K_{i4}\right]_{E,D6} + \left[K_{i4}\right]_{E,D15} + \left[K_{i4}\right]_{E,D18}$$

$$+ \left[K_{i1}\right]_{F,T5} + \left[K_{i1}\right]_{F,T9} + \left[K_{i1}\right]_{F,S9} + \left[K_{i1}\right]_{F,D11} + \left[K_{i2}\right]_{F,T7} + \left[K_{i2}\right]_{F,T8} + \left[K_{i2}\right]_{F,S3} + \left[K_{i2}\right]_{F,S4}$$

$$+ \left[K_{i2}\right]_{F,S7} + \left[K_{i2}\right]_{F,S8} + \left[K_{i2}\right]_{F,D9} + \left[K_{i2}\right]_{F,D12} + \left[K_{i2}\right]_{F,D23} + \left[K_{i3}\right]_{F,T6} + \left[K_{i3}\right]_{F,S2}$$

$$+ \left[K_{i3}\right]_{F,S5} + \left[K_{i3}\right]_{F,D5} + \left[K_{i3}\right]_{F,D6} + \left[K_{i3}\right]_{F,D15} + \left[K_{i3}\right]_{F,D18} + \left[K_{i3}\right]_{F,D24} + \left[K_{i1}\right]_{G,T7}$$

$$+ \left[K_{i2}\right]_{G,S5} + \left[K_{i2}\right]_{G,S12} + \left[K_{i2}\right]_{G,D5} + \left[K_{i2}\right]_{G,D10} + \left[K_{i2}\right]_{G,D18} + \left[K_{i2}\right]_{G,D20} + \left[K_{i2}\right]_{G,D22}$$

$$+ \left[K_{i3}\right]_{G,T3} + \left[K_{i3}\right]_{G,T5} + \left[K_{i3}\right]_{G,S9} + \left[K_{i3}\right]_{G,S11} + \left[K_{i3}\right]_{G,D4} + \left[K_{i3}\right]_{G,D7} + \left[K_{i3}\right]_{G,D11}$$

$$+ \left[K_{i3}\right]_{H,T9} + \left[K_{i4}\right]_{H,T6} + \left[K_{i4}\right]_{H,T7} + \left[K_{i4}\right]_{H,T10} + \left[K_{i4}\right]_{H,S2} + \left[K_{i4}\right]_{H,S5} + \left[K_{i4}\right]_{H,S10}$$

$$+ \left[K_{i3}\right]_{G,D17} + \left[K_{i3}\right]_{G,D19} + \left[K_{i4}\right]_{G,T2} + \left[K_{i4}\right]_{G,T4} + \left[K_{i4}\right]_{G,S4} + \left[K_{i4}\right]_{G,S6} + \left[K_{i4}\right]_{G,S7}$$

$$+ \left[K_{i4}\right]_{G,D8} + \left[K_{i4}\right]_{G,D9} + \left[K_{i4}\right]_{G,D21} + \left[K_{i4}\right]_{G,D23} + \left[K_{i4}\right]_{G,D24} + \left[K_{i1}\right]_{H,T3} + \left[K_{i1}\right]_{H,D7}$$

$$+ \left[K_{i3}\right]_{H,T9} + \left[K_{i4}\right]_{H,T6} + \left[K_{i4}\right]_{H,T7} + \left[K_{i4}\right]_{H,T10} + \left[K_{i4}\right]_{H,S2} + \left[K_{i4}\right]_{H,S5} + \left[K_{i4}\right]_{H,S10}$$

$$+ \left[K_{i4}\right]_{H,S11} + \left[K_{i4}\right]_{H,S12} + \left[K_{i4}\right]_{H,D1} + \left[K_{i4}\right]_{H,D4} + \left[K_{i4}\right]_{H,D16} + \left[K_{i4}\right]_{H,D17} + \left[K_{i4}\right]_{H,D19}$$

$$+ \left[K_{i4}\right]_{H,D22}$$

$$P_1 = \left[K_{i1}\right]_{A,T8} + \left[K_{i1}\right]_{A,S8} + \left[K_{i1}\right]_{A,D12} + \left[K_{i2}\right]_{A,T1} + \left[K_{i2}\right]_{A,S1} + \left[K_{i2}\right]_{A,D3} + \left[K_{i2}\right]_{A,D14}$$
$$+ \left[K_{i2}\right]_{A,D24} + \left[K_{i3}\right]_{A,D21} + \left[K_{i2}\right]_{D,D1} + \left[K_{i2}\right]_{D,D2} + \left[K_{i2}\right]_{D,D13} + \left[K_{i2}\right]_{D,D16}$$
$$+ \left[K_{i3}\right]_{D,T2} + \left[K_{i3}\right]_{D,T10} + \left[K_{i3}\right]_{D,S6} + \left[K_{i3}\right]_{D,S10} + \left[K_{i3}\right]_{D,D10} + \left[K_{i3}\right]_{D,D20} + \left[K_{i1}\right]_{E,T5}$$
$$+ \left[K_{i2}\right]_{E,S3} + \left[K_{i2}\right]_{E,S7} + \left[K_{i2}\right]_{E,D9} + \left[K_{i2}\right]_{E,D23} + \left[K_{i3}\right]_{E,T6} + \left[K_{i3}\right]_{E,S2} + \left[K_{i3}\right]_{E,D5}$$
$$+ \left[K_{i3}\right]_{E,D6} + \left[K_{i3}\right]_{E,D15} + \left[K_{i3}\right]_{E,D18} + \left[K_{i1}\right]_{H,T7} + \left[K_{i2}\right]_{H,S5} + \left[K_{i2}\right]_{H,S12}$$
$$+ \left[K_{i2}\right]_{H,D22} + \left[K_{i3}\right]_{H,T3} + \left[K_{i3}\right]_{H,S11} + \left[K_{i3}\right]_{H,D4} + \left[K_{i3}\right]_{H,D7} + \left[K_{i3}\right]_{H,D17} + \left[K_{i3}\right]_{H,D19}$$

$$P_2 = \left[K_{i2}\right]_{A,T9} + \left[K_{i2}\right]_{A,S9} + \left[K_{i2}\right]_{A,S11} + \left[K_{i2}\right]_{A,D2} + \left[K_{i2}\right]_{A,D11} + \left[K_{i2}\right]_{A,D13}$$
$$+ \left[K_{i2}\right]_{A,D21} + \left[K_{i3}\right]_{A,T10} + \left[K_{i3}\right]_{A,S10} + \left[K_{i3}\right]_{A,D10} + \left[K_{i3}\right]_{A,D20} + \left[K_{i4}\right]_{A,T8}$$
$$+ \left[K_{i4}\right]_{A,S8} + \left[K_{i4}\right]_{A,D3} + \left[K_{i4}\right]_{A,D12} + \left[K_{i4}\right]_{A,D14} + \left[K_{i2}\right]_{E,S12} + \left[K_{i2}\right]_{E,D5}$$
$$+ \left[K_{i2}\right]_{E,D18} + \left[K_{i2}\right]_{E,D22} + \left[K_{i3}\right]_{E,T3} + \left[K_{i3}\right]_{E,T5} + \left[K_{i3}\right]_{E,D4} + \left[K_{i3}\right]_{E,D7}$$
$$+ \left[K_{i3}\right]_{E,D17} + \left[K_{i4}\right]_{E,T4} + \left[K_{i4}\right]_{E,S7} + \left[K_{i4}\right]_{E,D8} + \left[K_{i4}\right]_{E,D9} + \left[K_{i4}\right]_{E,D23}$$

$$P_3 = \left[K_{i1}\right]_{A,D10} + \left[K_{i1}\right]_{A,D20} + \left[K_{i2}\right]_{A,T10} + \left[K_{i2}\right]_{A,S10} + \left[K_{i2}\right]_{A,D19} + \left[K_{i3}\right]_{A,D8}$$
$$+ \left[K_{i4}\right]_{A,T1} + \left[K_{i4}\right]_{A,S1} + \left[K_{i4}\right]_{A,D2} + \left[K_{i4}\right]_{A,D13} + \left[K_{i2}\right]_{B,D6} + \left[K_{i2}\right]_{B,D15}$$
$$+ \left[K_{i2}\right]_{B,D21} + \left[K_{i3}\right]_{B,T2} + \left[K_{i3}\right]_{B,S6} + \left[K_{i4}\right]_{B,T8} + \left[K_{i4}\right]_{B,S8} + \left[K_{i4}\right]_{B,D3}$$
$$+ \left[K_{i4}\right]_{B,D12} + \left[K_{i4}\right]_{B,D14} + \left[K_{i1}\right]_{E,T3} + \left[K_{i1}\right]_{E,D7} + \left[K_{i4}\right]_{E,T6} + \left[K_{i4}\right]_{E,S2}$$
$$+ \left[K_{i4}\right]_{E,S12} + \left[K_{i4}\right]_{E,D1} + \left[K_{i4}\right]_{E,D4} + \left[K_{i4}\right]_{E,D16} + \left[K_{i4}\right]_{E,D17} + \left[K_{i4}\right]_{E,D22}$$
$$+ \left[K_{i1}\right]_{F,T7} + \left[K_{i2}\right]_{F,S5} + \left[K_{i2}\right]_{F,D5} + \left[K_{i2}\right]_{F,D18} + \left[K_{i3}\right]_{F,T5} + \left[K_{i3}\right]_{F,S9}$$
$$+ \left[K_{i3}\right]_{F,D11} + \left[K_{i4}\right]_{F,S4} + \left[K_{i4}\right]_{F,S7} + \left[K_{i4}\right]_{F,D9} + \left[K_{i4}\right]_{F,D23} + \left[K_{i4}\right]_{F,D24}$$

$$P_4 = \left[K_{i1}\right]_{B,T4} + \left[K_{i2}\right]_{B,T2} + \left[K_{i2}\right]_{B,S6} + \left[K_{i3}\right]_{B,S4} + \left[K_{i4}\right]_{B,T1} + \left[K_{i4}\right]_{B,S1}$$
$$+ \left[K_{i4}\right]_{B,S3} + \left[K_{i3}\right]_{F,T9} + \left[K_{i4}\right]_{F,T6} + \left[K_{i4}\right]_{F,T7} + \left[K_{i4}\right]_{F,S2} + \left[K_{i4}\right]_{F,S5}$$

$$P_5 = \left[K_{i1}\right]_{B,T1} + \left[K_{i1}\right]_{B,S1} + \left[K_{i1}\right]_{B,D3} + \left[K_{i1}\right]_{B,D14} + \left[K_{i1}\right]_{B,D21} + \left[K_{i1}\right]_{B,D24}$$
$$+ \left[K_{i3}\right]_{B,T8} + \left[K_{i3}\right]_{B,S8} + \left[K_{i3}\right]_{B,D12} + \left[K_{i1}\right]_{C,D1} + \left[K_{i1}\right]_{C,D10} + \left[K_{i1}\right]_{C,D16}$$
$$+ \left[K_{i1}\right]_{C,D20} + \left[K_{i2}\right]_{C,T2} + \left[K_{i2}\right]_{C,T10} + \left[K_{i2}\right]_{C,S6} + \left[K_{i2}\right]_{C,S10}$$
$$+ \left[K_{i4}\right]_{C,D2} + \left[K_{i4}\right]_{C,D13} + \left[K_{i1}\right]_{F,T6} + \left[K_{i1}\right]_{F,S2} + \left[K_{i3}\right]_{F,S3}$$
$$+ \left[K_{i3}\right]_{F,S7} + \left[K_{i3}\right]_{F,D9} + \left[K_{i3}\right]_{F,D23} + \left[K_{i4}\right]_{F,T5} + \left[K_{i4}\right]_{F,D5} + \left[K_{i4}\right]_{F,D6}$$
$$+ \left[K_{i4}\right]_{F,D15} + \left[K_{i4}\right]_{F,D18} + \left[K_{i1}\right]_{G,T3} + \left[K_{i1}\right]_{G,D7} + \left[K_{i4}\right]_{G,T7} + \left[K_{i4}\right]_{G,S5}$$
$$+ \left[K_{i4}\right]_{G,S11} + \left[K_{i4}\right]_{G,S12} + \left[K_{i4}\right]_{G,D4} + \left[K_{i4}\right]_{G,D17} + \left[K_{i4}\right]_{G,D19} + \left[K_{i4}\right]_{G,D22}$$

$$P_6 = \left[K_{i1}\right]_{C,T10} + \left[K_{i1}\right]_{C,S10} + \left[K_{i1}\right]_{C,S11} + \left[K_{i1}\right]_{C,D2} + \left[K_{i1}\right]_{C,D3} + \left[K_{i1}\right]_{C,D13}$$
$$+ \left[K_{i1}\right]_{C,D14} + \left[K_{i1}\right]_{C,D21} + \left[K_{i3}\right]_{C,T8} + \left[K_{i3}\right]_{C,S8} + \left[K_{i3}\right]_{C,D12} + \left[K_{i4}\right]_{C,T9}$$
$$+ \left[K_{i4}\right]_{C,S9} + \left[K_{i4}\right]_{C,D10} + \left[K_{i4}\right]_{C,D11} + \left[K_{i4}\right]_{C,D20} + \left[K_{i1}\right]_{G,D4} + \left[K_{i1}\right]_{G,D17}$$
$$+ \left[K_{i2}\right]_{G,T3} + \left[K_{i2}\right]_{G,D7} + \left[K_{i2}\right]_{G,D8} + \left[K_{i3}\right]_{G,T4} + \left[K_{i3}\right]_{G,S7} + \left[K_{i3}\right]_{G,S12}$$
$$+ \left[K_{i3}\right]_{G,D9} + \left[K_{i3}\right]_{G,D22} + \left[K_{i3}\right]_{G,D23} + \left[K_{i4}\right]_{G,T5} + \left[K_{i4}\right]_{G,D5} + \left[K_{i4}\right]_{G,D18}$$

$$P_7 = \left[K_{i1}\right]_{C,T2} + \left[K_{i1}\right]_{C,T8} + \left[K_{i1}\right]_{C,S6} + \left[K_{i1}\right]_{C,S8} + \left[K_{i1}\right]_{C,D6} + \left[K_{i1}\right]_{C,D12}$$
$$+ \left[K_{i1}\right]_{C,D15} + \left[K_{i2}\right]_{C,D3} + \left[K_{i2}\right]_{C,D14} + \left[K_{i3}\right]_{C,D21} + \left[K_{i1}\right]_{D,T1} + \left[K_{i1}\right]_{D,T10}$$
$$+ \left[K_{i1}\right]_{D,S1} + \left[K_{i1}\right]_{D,S10} + \left[K_{i1}\right]_{D,D2} + \left[K_{i1}\right]_{D,D8} + \left[K_{i1}\right]_{D,D13} + \left[K_{i1}\right]_{D,D19}$$
$$+ \left[K_{i4}\right]_{D,D10} + \left[K_{i4}\right]_{D,D20} + \left[K_{i1}\right]_{G,T5} + \left[K_{i1}\right]_{G,S9} + \left[K_{i1}\right]_{G,D11} + \left[K_{i2}\right]_{G,T7}$$
$$+ \left[K_{i2}\right]_{G,S4} + \left[K_{i2}\right]_{G,S7} + \left[K_{i2}\right]_{G,D9} + \left[K_{i2}\right]_{G,D23} + \left[K_{i3}\right]_{G,S5} + \left[K_{i3}\right]_{G,D5}$$
$$+ \left[K_{i3}\right]_{G,D18} + \left[K_{i3}\right]_{G,D24} + \left[K_{i1}\right]_{H,T6} + \left[K_{i1}\right]_{H,S2} + \left[K_{i1}\right]_{H,D4} + \left[K_{i1}\right]_{H,D17}$$
$$+ \left[K_{i2}\right]_{H,T3} + \left[K_{i2}\right]_{H,D7} + \left[K_{i3}\right]_{H,S12} + \left[K_{i3}\right]_{H,D1} + \left[K_{i3}\right]_{H,D16} + \left[K_{i3}\right]_{H,D22}$$

$$P_8 = \left[K_{i1}\right]_{D,T2} + \left[K_{i1}\right]_{D,S3} + \left[K_{i1}\right]_{D,S4} + \left[K_{i1}\right]_{D,S6} + \left[K_{i2}\right]_{D,T1} + \left[K_{i2}\right]_{D,T4}$$
$$+ \left[K_{i2}\right]_{D,S1} + \left[K_{i1}\right]_{H,T9} + \left[K_{i2}\right]_{H,T7} + \left[K_{i3}\right]_{H,T6} + \left[K_{i3}\right]_{H,S2} + \left[K_{i3}\right]_{H,S5}$$

$$P_9 = \left[K_{i1}\right]_{A,T9} + \left[K_{i1}\right]_{A,S9} + \left[K_{i1}\right]_{A,D11} + \left[K_{i2}\right]_{A,T8} + \left[K_{i2}\right]_{A,S7} + \left[K_{i2}\right]_{A,S8}$$
$$+ \left[K_{i2}\right]_{A,D9} + \left[K_{i2}\right]_{A,D12} + \left[K_{i2}\right]_{A,D23} + \left[K_{i3}\right]_{A,T6} + \left[K_{i3}\right]_{A,D24} + \left[K_{i2}\right]_{D,S5}$$
$$+ \left[K_{i2}\right]_{D,D10} + \left[K_{i2}\right]_{D,D20} + \left[K_{i3}\right]_{D,T3} + \left[K_{i3}\right]_{D,D7} + \left[K_{i3}\right]_{D,D19} + \left[K_{i4}\right]_{D,T2}$$
$$+ \left[K_{i4}\right]_{D,T4} + \left[K_{i4}\right]_{D,S4} + \left[K_{i4}\right]_{D,S6} + \left[K_{i4}\right]_{D,D8}$$

$$P_{10} = \left[K_{i2}\right]_{A,S12} + \left[K_{i2}\right]_{A,D10} + \left[K_{i2}\right]_{A,D20} + \left[K_{i2}\right]_{A,D22} + \left[K_{i3}\right]_{A,S9} + \left[K_{i3}\right]_{A,S11}$$
$$+ \left[K_{i3}\right]_{A,D11} + \left[K_{i3}\right]_{A,D19} + \left[K_{i4}\right]_{A,S7} + \left[K_{i4}\right]_{A,D8} + \left[K_{i4}\right]_{A,D9} + \left[K_{i4}\right]_{A,D21}$$
$$+ \left[K_{i4}\right]_{A,D23} + \left[K_{i4}\right]_{A,D24}$$

$$P_{11} = \left[K_{i3}\right]_{A,T9} + \left[K_{i4}\right]_{A,T6} + \left[K_{i4}\right]_{A,T10} + \left[K_{i4}\right]_{A,S10} + \left[K_{i4}\right]_{A,S11} + \left[K_{i4}\right]_{A,S12}$$
$$+ \left[K_{i4}\right]_{A,D19} + \left[K_{i4}\right]_{A,D22} + \left[K_{i3}\right]_{B,T5} + \left[K_{i4}\right]_{B,T2} + \left[K_{i4}\right]_{B,T4} + \left[K_{i4}\right]_{B,S4}$$
$$+ \left[K_{i4}\right]_{B,S6} + \left[K_{i4}\right]_{B,D21} + \left[K_{i4}\right]_{B,D24}$$

$$P_{12} = 0$$

$$P_{13} = \left[K_{i3}\right]_{B,T1} + \left[K_{i3}\right]_{B,T4} + \left[K_{i3}\right]_{B,S1} + \left[K_{i3}\right]_{B,S3} + \left[K_{i3}\right]_{B,D3} + \left[K_{i3}\right]_{B,D14}$$
$$+ \left[K_{i4}\right]_{B,T5} + \left[K_{i4}\right]_{B,D6} + \left[K_{i4}\right]_{B,D15} + \left[K_{i3}\right]_{C,T9} + \left[K_{i4}\right]_{C,T7} + \left[K_{i4}\right]_{C,T10}$$
$$+ \left[K_{i4}\right]_{C,S10} + \left[K_{i4}\right]_{C,S11} + \left[K_{i4}\right]_{C,D1} + \left[K_{i4}\right]_{C,D4} + \left[K_{i4}\right]_{C,D16} + \left[K_{i4}\right]_{C,D17}$$

$$P_{14} = \left[K_{i1}\right]_{C,D4} + \left[K_{i1}\right]_{C,D17} + \left[K_{i3}\right]_{C,D1} + \left[K_{i3}\right]_{C,D2} + \left[K_{i3}\right]_{C,D3} + \left[K_{i3}\right]_{C,D13}$$
$$+ \left[K_{i3}\right]_{C,D14} + \left[K_{i3}\right]_{C,D16} + \left[K_{i4}\right]_{C,D5} + \left[K_{i4}\right]_{C,D6} + \left[K_{i4}\right]_{C,D15} + \left[K_{i4}\right]_{C,D18}$$

$$P_{15} = \left[K_{i1}\right]_{C,T9} + \left[K_{i1}\right]_{C,S9} + \left[K_{i1}\right]_{C,D11} + \left[K_{i2}\right]_{C,T7} + \left[K_{i2}\right]_{C,T8} + \left[K_{i2}\right]_{C,S8}$$
$$+ \left[K_{i2}\right]_{C,D12} + \left[K_{i3}\right]_{C,D5} + \left[K_{i3}\right]_{C,D6} + \left[K_{i3}\right]_{C,D15} + \left[K_{i3}\right]_{C,D18} + \left[K_{i1}\right]_{D,S2}$$
$$+ \left[K_{i2}\right]_{D,T3} + \left[K_{i2}\right]_{D,D7} + \left[K_{i2}\right]_{D,D8} + \left[K_{i3}\right]_{D,T1} + \left[K_{i3}\right]_{D,T4} + \left[K_{i3}\right]_{D,S1}$$
$$+ \left[K_{i3}\right]_{D,S3} + \left[K_{i3}\right]_{D,D1} + \left[K_{i3}\right]_{D,D2} + \left[K_{i3}\right]_{D,D13} + \left[K_{i3}\right]_{D,D16}$$

$$P_{16} = \left[K_{i2}\right]_{D,S4} + \left[K_{i2}\right]_{D,S3} + \left[K_{i3}\right]_{D,S2} + \left[K_{i3}\right]_{D,S5}$$

$$P_{17} = \left[K_{i1}\right]_{E,S3} + \left[K_{i1}\right]_{E,D6} + \left[K_{i1}\right]_{E,D15} + \left[K_{i2}\right]_{E,T1} + \left[K_{i2}\right]_{E,T4} + \left[K_{i2}\right]_{E,T5}$$
$$+ \left[K_{i2}\right]_{E,S1} + \left[K_{i2}\right]_{E,D3} + \left[K_{i2}\right]_{E,D14} + \left[K_{i2}\right]_{H,T9} + \left[K_{i2}\right]_{H,S11} + \left[K_{i2}\right]_{H,D1}$$
$$+ \left[K_{i2}\right]_{H,D4} + \left[K_{i2}\right]_{H,D16} + \left[K_{i2}\right]_{H,D17} + \left[K_{i3}\right]_{H,T7} + \left[K_{i3}\right]_{H,T10} + \left[K_{i3}\right]_{H,S10}$$

$$P_{18} = \left[K_{i1}\right]_{E,D5} + \left[K_{i1}\right]_{E,D18} + \left[K_{i2}\right]_{E,D1} + \left[K_{i2}\right]_{E,D2} + \left[K_{i2}\right]_{E,D4} + \left[K_{i2}\right]_{E,D6}$$
$$+ \left[K_{i2}\right]_{E,D13} + \left[K_{i2}\right]_{E,D15} + \left[K_{i2}\right]_{E,D16} + \left[K_{i2}\right]_{E,D17} + \left[K_{i4}\right]_{E,D3} + \left[K_{i4}\right]_{E,D14}$$

$$P_{19} = \left[K_{i1}\right]_{E,T4} + \left[K_{i1}\right]_{E,D1} + \left[K_{i1}\right]_{E,D16} + \left[K_{i2}\right]_{E,S2} + \left[K_{i3}\right]_{E,D8} + \left[K_{i4}\right]_{E,T1}$$
$$+ \left[K_{i4}\right]_{E,T3} + \left[K_{i4}\right]_{E,S1} + \left[K_{i4}\right]_{E,S3} + \left[K_{i4}\right]_{E,D2} + \left[K_{i4}\right]_{E,D7} + \left[K_{i4}\right]_{E,D13}$$
$$+ \left[K_{i1}\right]_{F,D5} + \left[K_{i1}\right]_{F,D18} + \left[K_{i2}\right]_{F,T9} + \left[K_{i2}\right]_{F,S9} + \left[K_{i2}\right]_{F,D6} + \left[K_{i2}\right]_{F,D11}$$
$$+ \left[K_{i2}\right]_{F,D15} + \left[K_{i3}\right]_{F,T7} + \left[K_{i4}\right]_{F,T8} + \left[K_{i4}\right]_{F,S8} + \left[K_{i4}\right]_{F,D12}$$

$$P_{20} = \left[K_{i1}\right]_{F,S5} + \left[K_{i2}\right]_{F,S2} + \left[K_{i3}\right]_{F,S4} + \left[K_{i4}\right]_{F,S3}$$

$$P_{21} = \left[K_{i1}\right]_{F,S7} + \left[K_{i1}\right]_{F,D9} + \left[K_{i1}\right]_{F,D23} + \left[K_{i1}\right]_{F,D24} + \left[K_{i2}\right]_{F,T6} + \left[K_{i3}\right]_{F,T8}$$
$$+ \left[K_{i3}\right]_{F,S8} + \left[K_{i3}\right]_{F,D12} + \left[K_{i4}\right]_{F,T9} + \left[K_{i4}\right]_{F,S9} + \left[K_{i4}\right]_{F,D11} + \left[K_{i1}\right]_{G,T4}$$
$$+ \left[K_{i1}\right]_{G,S5} + \left[K_{i1}\right]_{G,D10} + \left[K_{i1}\right]_{G,D20} + \left[K_{i2}\right]_{G,T2} + \left[K_{i2}\right]_{G,S6} + \left[K_{i2}\right]_{G,D19}$$
$$+ \left[K_{i3}\right]_{G,S4} + \left[K_{i3}\right]_{G,D8} + \left[K_{i4}\right]_{G,T3} + \left[K_{i4}\right]_{G,D7}$$

$$P_{22} = \left[K_{i1}\right]_{G,S7} + \left[K_{i1}\right]_{G,S11} + \left[K_{i1}\right]_{G,S12} + \left[K_{i1}\right]_{G,D8} + \left[K_{i1}\right]_{G,D9} + \left[K_{i1}\right]_{G,D19} + \left[K_{i1}\right]_{G,D21}$$
$$+ \left[K_{i1}\right]_{G,D22} + \left[K_{i1}\right]_{G,D23} + \left[K_{i1}\right]_{G,D24} + \left[K_{i4}\right]_{G,S9} + \left[K_{i4}\right]_{G,D10} + \left[K_{i4}\right]_{G,D11} + \left[K_{i4}\right]_{G,D20}$$

$$P_{23} = \left[K_{i1}\right]_{G,T2} + \left[K_{i1}\right]_{G,S4} + \left[K_{i1}\right]_{G,S6} + \left[K_{i2}\right]_{G,T4} + \left[K_{i2}\right]_{G,T5} + \left[K_{i2}\right]_{G,D24}$$
$$+ \left[K_{i3}\right]_{G,D21} + \left[K_{i1}\right]_{H,T10} + \left[K_{i1}\right]_{H,S10} + \left[K_{i1}\right]_{H,S11} + \left[K_{i1}\right]_{H,S12} + \left[K_{i1}\right]_{H,D19}$$
$$+ \left[K_{i1}\right]_{H,D22} + \left[K_{i2}\right]_{H,T6} + \left[K_{i4}\right]_{H,T9}$$

$$P_{24} = 0$$

$$P_{25} = \left[K_{i1}\right]_{A,T6} + \left[K_{i2}\right]_{A,D8} + \left[K_{i3}\right]_{A,T1} + \left[K_{i3}\right]_{A,S1} + \left[K_{i3}\right]_{A,S7} + \left[K_{i3}\right]_{A,S12}$$
$$+ \left[K_{i3}\right]_{A,D2} + \left[K_{i3}\right]_{A,D3} + \left[K_{i3}\right]_{A,D9} + \left[K_{i3}\right]_{A,D13} + \left[K_{i3}\right]_{A,D14} + \left[K_{i3}\right]_{A,D22}$$
$$+ \left[K_{i3}\right]_{A,D23} + \left[K_{i1}\right]_{B,T5} + \left[K_{i2}\right]_{B,T8} + \left[K_{i2}\right]_{B,S3} + \left[K_{i2}\right]_{B,S4} + \left[K_{i2}\right]_{B,S8}$$
$$+ \left[K_{i3}\right]_{B,D12} + \left[K_{i3}\right]_{B,D6} + \left[K_{i3}\right]_{B,D15} + \left[K_{i3}\right]_{B,D24} + \left[K_{i1}\right]_{C,T7} + \left[K_{i2}\right]_{C,D5}$$
$$+ \left[K_{i2}\right]_{C,D10} + \left[K_{i2}\right]_{C,D18} + \left[K_{i2}\right]_{C,D20} + \left[K_{i3}\right]_{C,S9} + \left[K_{i3}\right]_{C,S11} + \left[K_{i3}\right]_{C,D4}$$
$$+ \left[K_{i3}\right]_{C,D11} + \left[K_{i3}\right]_{C,D17} + \left[K_{i4}\right]_{C,T2} + \left[K_{i4}\right]_{C,S6} + \left[K_{i4}\right]_{C,D21} + \left[K_{i1}\right]_{D,T3}$$
$$+ \left[K_{i1}\right]_{D,D7} + \left[K_{i4}\right]_{D,T10} + \left[K_{i4}\right]_{D,S2} + \left[K_{i4}\right]_{D,S5} + \left[K_{i4}\right]_{D,S10} + \left[K_{i4}\right]_{D,D1}$$
$$+ \left[K_{i4}\right]_{D,D16} + \left[K_{i4}\right]_{D,D19}$$

$$P_{26} = \left[K_{i1}\right]_{E,T1} + \left[K_{i1}\right]_{E,S1} + \left[K_{i1}\right]_{E,S7} + \left[K_{i1}\right]_{E,S12} + \left[K_{i1}\right]_{E,D2} + \left[K_{i1}\right]_{E,D3}$$
$$+ \left[K_{i1}\right]_{E,D8} + \left[K_{i1}\right]_{E,D9} + \left[K_{i1}\right]_{E,D13} + \left[K_{i1}\right]_{E,D14} + \left[K_{i1}\right]_{E,D22} + \left[K_{i1}\right]_{E,D23}$$
$$+ \left[K_{i2}\right]_{E,T6} + \left[K_{i1}\right]_{F,T8} + \left[K_{i1}\right]_{F,S3} + \left[K_{i1}\right]_{F,S4} + \left[K_{i1}\right]_{F,S8} + \left[K_{i1}\right]_{F,D6}$$
$$+ \left[K_{i1}\right]_{F,D12} + \left[K_{i1}\right]_{F,D15} + \left[K_{i1}\right]_{F,D24} + \left[K_{i2}\right]_{F,T5} + \left[K_{i1}\right]_{G,D5} + \left[K_{i1}\right]_{G,D18}$$
$$+ \left[K_{i2}\right]_{G,S9} + \left[K_{i2}\right]_{G,S11} + \left[K_{i2}\right]_{G,D4} + \left[K_{i2}\right]_{G,D11} + \left[K_{i2}\right]_{G,D17} + \left[K_{i2}\right]_{G,D21}$$
$$+ \left[K_{i3}\right]_{G,T2} + \left[K_{i3}\right]_{G,T7} + \left[K_{i3}\right]_{G,S6} + \left[K_{i3}\right]_{G,D10} + \left[K_{i3}\right]_{G,D20} + \left[K_{i1}\right]_{H,S5}$$
$$+ \left[K_{i1}\right]_{H,D1} + \left[K_{i1}\right]_{H,D16} + \left[K_{i2}\right]_{H,T10} + \left[K_{i2}\right]_{H,S2} + \left[K_{i2}\right]_{H,S10} + \left[K_{i2}\right]_{H,D19}$$
$$+ \left[K_{i4}\right]_{H,T3} + \left[K_{i3}\right]_{H,D7}$$

So long as mesh nodes lie on the boundary interface plane, suitable tetrahedral element subdivision can be made automatically, requiring only brick mesh node data and knowledge of the boundary surface. Once this information is programmed, it is possible to model a wide variety of different boundary shapes in three dimensions.

Chapter 7
FEM FORMULATION IN MAGNETOSTATICS

The finite element formulation for non-saturated magnetic problems using the magnetic scalar potential ϕ in two dimensions is in every respect identical to its electric scalar potential counterpart. The only difference is that the dielectric permittivity is replaced with the magnetic permeability. For the variational approach, the functional to be minimised is equal to the total magnetic energy W

$$W = \frac{1}{2}\int_{\Omega} \mathbf{H} \cdot \mathbf{B} \, d\Omega = \frac{1}{2}\int_{\Omega} \mu |\nabla \phi|^2 \, d\Omega \qquad (7.1)$$

where Ω is the total domain. It should be noted that using the above energy expression assumes that the potential is either fixed on the outer boundary or that its normal derivative is zero. For more general boundary conditions, the weak formulation can be used. The boundary conditions on the scalar potential are not trivial to derive. The scalar potential must not span current carrying regions and care should be taken in defining the domain so as to avoid multi-valued boundary conditions. Note that for the axi-symmetric case, where first-order elements are used, corrective action for the nodes on-axis must be taken, just as in the electrostatic case.

1.   MAGNETIC VECTOR POTENTIAL

1.1   Two dimensional field distributions

For the vector potential, the following functional must be used

$$F = \int_{\Omega} (W - \mathbf{J} \cdot \mathbf{A}) \, d\Omega \qquad (7.2)$$

This functional is obviously similar to the functional corresponding to Poisson's equation, with the difference that the source term is derived from the scalar product of two vectors. For no saturation effects, this functional can obviously be expressed in terms of the vector potential **A** in the following way

$$F = \int_\Omega \left(\frac{1}{2\mu} |\nabla \times \mathbf{A}|^2 - \mathbf{J} \cdot \mathbf{A} \right) d\Omega \tag{7.3}$$

For the situation where the source is infinitely long and points in the z-direction only, then the functional simplifies to

$$F = \int_\Omega \left(\frac{1}{2\mu} \left(\left(\frac{\partial A}{\partial x} \right)^2 + \left(\frac{\partial A}{\partial y} \right)^2 \right) - \mathbf{J} \cdot \mathbf{A} \right) dxdy \tag{7.4}$$

Both J(x,y), the known source current distribution and A(x,y), the vector potential distribution point in the z-direction. This functional is similar to the one for Poisson's equation in electrostatics and its finite element solution proceeds in exactly the same way. The outer boundaries are typically extended as far as possible from the region of interest.

For the rotationally symmetric structure, where the source current J(r,z,) and the vector potential A(r,z) point in the azimuthal direction ϕ, then the functional for non-saturated problems takes the following form

$$F = 2\pi \iint_\Omega \left\{ \frac{1}{2\mu} \left[\left(\frac{\partial A}{\partial z} \right)^2 + \left(\frac{A}{r} + \frac{\partial A}{\partial r} \right)^2 \right] - JA \right\} rdrdz \tag{7.5}$$

Expanding A in terms of the shape functions for m surrounding elements and differentiating F with respect to node i gives

$$\left[\sum_j K_{ij} u_j + G_i \right]_{A_{e1}} + \left[\sum_j K_{ij} u_j + G_i \right]_{A_{e2}} + \ldots\ldots\ldots \left[\sum_j K_{ij} u_j + G_i \right]_{A_{em}} = 0$$

where

$$K_{ij} = 2\pi \iint_{\Omega_e} \frac{1}{\mu} \left[\left(\frac{\partial N_i}{\partial z} \right) \left(\frac{\partial N_j}{\partial z} \right) + \left(\frac{N_i}{r} + \frac{\partial N_i}{\partial r} \right) \left(\frac{N_j}{r} + \frac{\partial N_j}{\partial r} \right) \right] rdrdz$$

$$G_i = -2\pi \iint_{\Omega_e} JN_i rdrdz \tag{7.6}$$

These expressions can be accurately and efficiently integrated numerically once the type and order of the shape to be used is decided upon. The value of the vector potential on axis is specified to be zero, since it is an odd function (given that the flux density must be an even function).

It should be noted that the singular term A/r occurs in the functional F, and this propagates through to a singular term in the stiffness matrix entries. Although on-axis nodes do not occur, the nodes near the optic axis will be effected. A singular expression such as this can be avoided if the transformation P=2A/r is used, giving the following

$$K_{ij} = \pi \iint_{\Omega_e} \frac{1}{\mu} \left[r^2 \left(\frac{\partial N_i}{\partial z} \right) \left(\frac{\partial N_j}{\partial z} \right) + \left(2N_i + r \frac{\partial N_i}{\partial r} \right) \left(2N_j + r \frac{\partial N_j}{\partial r} \right) \right] r dr dz$$

$$G_i = -\pi \iint_{\Omega_e} JN_i r^2 dr dz$$

(7.7)

Not only does this avoid the singular term, the function P(r,z) on the axis (r=0) corresponds directly to the flux density distribution (equation 1.67), which is the main parameter that is used for the determination of the first-order optics and on-axis aberration coefficients of a magnetic electron lens. It should be noted that if first-order elements are used, the nodes on the axis must be corrected for their radial variation in much the same way that is required for the scalar potential formulation in axi-symmetric systems.

Another way of reducing the effect of the singular term is to use u=rA. In this case

$$K_{ij} = 2\pi \iint_{\Omega_e} \frac{1}{\mu r} \left[\left(\frac{\partial N_i}{\partial z} \right) \left(\frac{\partial N_j}{\partial z} \right) + \left(\frac{\partial N_i}{\partial r} \right) \left(\frac{\partial N_j}{\partial r} \right) \right] dr dz$$

$$G_i = -2\pi \iint_{\Omega_e} JN_i dr dz$$

(7.8)

Although this formulation does not eliminate the singularity, it does make it symmetric with respect to the r and z directions.

1.2   Three dimensional field distributions

The vector potential formulation in the three dimensions (x,y,z) proceeds as before, but now solves for the three components, $A_x(x,y,z)$, $A_y(x,y,z)$ and $A_z(x,y,z)$. Firstly, in terms of the three components of the flux density $B_x(x,y,z)$, $B_y(x,y,z)$ and $B_z(x,y,z)$, the functional becomes

$$F = \int_\Omega \left(\frac{1}{2\mu} \left(B_x{}^2 + B_y{}^2 + B_z{}^2 \right) - \left(J_x A_x + J_y A_y + J_z A_z \right) \right) dxdydz \quad (7.9)$$

where $J_x(x,y,z)$, $J_y(x,y,z)$ and $J_z(x,y,z)$ are known current distributions. The three components of $\mathbf{A}$ can be expanded in terms of shape functions

$$A_x = \sum_j N_j A_{xj} \quad A_y = \sum_j N_j A_{yj} \quad A_z = \sum_j N_j A_{zj} \quad (7.10)$$

The function F is then minimised with respect to each of these functions at the node i

$$\frac{\partial F}{\partial A_{xi}} = \int_\Omega \left(\frac{1}{\mu} \left(B_y \frac{\partial B_y}{\partial A_{xi}} + B_z \frac{\partial B_z}{\partial A_{xi}} \right) - \left(J_x N_i \right) \right) dxdydz$$

$$\frac{\partial F}{\partial A_{yi}} = \int_\Omega \left(\frac{1}{\mu} \left(B_x \frac{\partial B_x}{\partial A_{yi}} + B_z \frac{\partial B_z}{\partial A_{yi}} \right) - \left(J_y N_i \right) \right) dxdydz \quad (7.11)$$

$$\frac{\partial F}{\partial A_{zi}} = \int_\Omega \left(\frac{1}{\mu} \left(B_y \frac{\partial B_y}{\partial A_{zi}} + B_x \frac{\partial B_x}{\partial A_{zi}} \right) - \left(J_z N_i \right) \right) dxdydz$$

These equations can obviously be further simplified to generate matrix equations involving the unknown values of A_x, A_y and A_z at the mesh nodes.

In this variational approach, it is not clear what is assumed about the boundary conditions. For this reason, it is instructive to consider the more general weighted residual technique. Being a vector equation, the initial differential equation has three components, so applying the weighted residual method to it must involve using three weighting functions. Let each of these weighting functions be N_i, where i runs from 1 to n and denotes the number of unknown nodes in the domain. The resulting residual $\mathbf{R}_i$ is a vector, having three components (R_{xi}, R_{yi}, R_{zi}). Each weighted equation is formed from equation (1.90) and is thus given by

$$R_{pi} = \int_\Omega N_i \, \hat{p} \cdot \left[\nabla \times \left(\frac{1}{\mu} \nabla \times \mathbf{A} \right) - \mathbf{J} \right] d\Omega = 0 \quad (7.12)$$

where $\hat{p}$ denotes the x, y and z directions.

To simplify each component equation, we can make use of Green's first vector theorem (given in Appendix 3), so

$$\int_{\Omega} N_i \hat{p} \cdot \left(\nabla \times \frac{1}{\mu} \nabla \times \mathbf{A} \right) d\Omega = \int_{\Omega} \frac{1}{\mu} \left(\nabla \times N_i \hat{p} \right) \cdot \left(\nabla \times \mathbf{A} \right) d\Omega - \int_S \frac{1}{\mu} \left(N_i \hat{p} \times \nabla \times \mathbf{A} \right) \cdot \mathbf{n} dS$$

$$(7.13)$$

The surface boundary S encloses the volume Ω. The surface term on the right-hand side of this equation is zero when all three components of the vector potential at the surface are specified explicitly: in this case no node i lies on the surface, so N_i will always be zero on the boundary. On the other hand, the surface term can be expressed in terms of the flux density **B**, and using the vector theorem that $\mathbf{a} \cdot (\mathbf{b} \times \mathbf{c}) = \mathbf{b} \cdot (\mathbf{c} \times \mathbf{a}) = \mathbf{c} \cdot (\mathbf{a} \times \mathbf{b})$, then it can be rearranged to

$$\int_S \frac{1}{\mu} \left(N_i \hat{p} \times \nabla \times \mathbf{A} \right) \cdot \mathbf{n} dS = \int_S \frac{1}{\mu} (\mathbf{n} \times \mathbf{B}) \cdot \left(N_i \hat{p} \right) dS \qquad (7.14)$$

If the direction of the flux is normal to the outer boundary, where $(\mathbf{n} \times \mathbf{B})dS$ is zero, then the surface integral term will also be zero and can be left out of the final matrix equation. As in the scalar case, if it is left out, and the nodes on the boundary are specified to be free nodes, then the normal flux condition at the boundary will automatically be enforced.

The residual equations are expressed as

$$R_{xi} = \int_{\Omega} \frac{1}{\mu} \left(\nabla \times N_{ix} \right) \cdot \left(\nabla \times \mathbf{A} \right) - N_{ix} J_x d\Omega = 0$$

$$R_{yi} = \int_{\Omega} \frac{1}{\mu} \left(\nabla \times N_{iy} \right) \cdot \left(\nabla \times \mathbf{A} \right) - N_{iy} J_y d\Omega = 0$$

$$R_{zi} = \int_{\Omega} \frac{1}{\mu} \left(\nabla \times N_{iz} \right) \cdot \left(\nabla \times \mathbf{A} \right) - N_{iz} J_z d\Omega = 0 \qquad (7.15)$$

We have

$$\left(\nabla \times \mathbf{A} \right) = \left(\frac{\partial A_z}{\partial y} - \frac{\partial A_y}{\partial z} \right) \hat{x} - \left(\frac{\partial A_z}{\partial x} - \frac{\partial A_x}{\partial z} \right) \hat{y} + \left(\frac{\partial A_y}{\partial x} - \frac{\partial A_x}{\partial y} \right) \hat{z} \qquad (7.16)$$

and

$$\left(\nabla \times N_{ix} \right) = \left(\frac{\partial N_{ix}}{\partial z} \right) \hat{y} - \left(\frac{\partial N_{ix}}{\partial y} \right) \hat{z} \qquad \left(\nabla \times N_{iy} \right) = -\left(\frac{\partial N_{iy}}{\partial z} \right) \hat{x} + \left(\frac{\partial N_{iy}}{\partial x} \right) \hat{z}$$

$$\left(\nabla \times N_{iz} \right) = \left(\frac{\partial N_{iz}}{\partial y} \right) \hat{x} - \left(\frac{\partial N_{iz}}{\partial x} \right) \hat{y} \qquad (7.17)$$

The three residual equations corresponding to the x, y and z components can then be expressed as

$$R_{xi} = \sum_j \left(\int_\Omega \frac{1}{\mu} \left(\frac{\partial N_{ix}}{\partial z} \frac{\partial N_j}{\partial z} + \frac{\partial N_{ix}}{\partial y} \frac{\partial N_j}{\partial y} \right) A_{xj} - \frac{1}{\mu} \frac{\partial N_{ix}}{\partial y} \frac{\partial N_j}{\partial x} A_{yj} - \frac{1}{\mu} \frac{\partial N_{ix}}{\partial z} \frac{\partial N_j}{dx} A_{zj} d\Omega \right)$$
$$- \int_\Omega N_{ix} J_x d\Omega$$

$$R_{yi} = \sum_j \left(\int_\Omega - \frac{1}{\mu} \frac{\partial N_{iy}}{\partial x} \frac{\partial N_j}{\partial y} A_{xj} + \frac{1}{\mu} \left(\frac{\partial N_{iy}}{\partial x} \frac{\partial N_j}{\partial x} + \frac{\partial N_{iy}}{\partial z} \frac{\partial N_j}{\partial z} \right) A_{yj} - \frac{1}{\mu} \frac{\partial N_{yx}}{\partial z} \frac{\partial N_j}{dy} A_{zj} d\Omega \right)$$
$$- \int_v N_{iy} J_y d\Omega$$

$$R_{zi} = \sum_j \left(\int_\Omega - \frac{1}{\mu} \frac{\partial N_{iz}}{\partial x} \frac{\partial N_j}{\partial z} A_{xj} - \frac{1}{\mu} \frac{\partial N_{iz}}{\partial y} \frac{\partial N_j}{dz} A_{yj} + \frac{1}{\mu} \left(\frac{\partial N_{iz}}{\partial y} \frac{\partial N_j}{\partial y} + \frac{\partial N_{iz}}{\partial x} \frac{\partial N_j}{\partial x} \right) A_{zj} d\Omega \right)$$
$$- \int_v N_{iz} J_z d\Omega$$

$$(7.18)$$

Put in matrix form,

$$\begin{bmatrix} \mathbf{K}_{xx} & \mathbf{K}_{xy} & \mathbf{K}_{xz} \\ \mathbf{K}_{yx} & \mathbf{K}_{yy} & \mathbf{K}_{yz} \\ \mathbf{K}_{zx} & \mathbf{K}_{zy} & \mathbf{K}_{zz} \end{bmatrix} \begin{bmatrix} \mathbf{A}_x \\ \mathbf{A}_y \\ \mathbf{A}_z \end{bmatrix} = \begin{bmatrix} \mathbf{b}_x \\ \mathbf{b}_y \\ \mathbf{b}_z \end{bmatrix}$$

where the sub-matrices, $\mathbf{K}_{xx}$ to $\mathbf{K}_{zz}$ are of rank n by n, and the column vectors $\mathbf{A}_x$ through to $\mathbf{b}_z$ have rank 1 by n. The entries in these matrices and vectors are given by

$$K_{xx} = \int_\Omega \frac{1}{\mu} \left(\frac{\partial N_i}{\partial y} \frac{\partial N_j}{\partial y} + \frac{\partial N_i}{\partial z} \frac{\partial N_j}{\partial z} \right) d\Omega \qquad K_{yy} = \int_\Omega \frac{1}{\mu} \left(\frac{\partial N_i}{\partial z} \frac{\partial N_j}{\partial z} + \frac{\partial N_i}{\partial x} \frac{\partial N_j}{\partial x} \right) d\Omega$$

$$K_{zz} = \int_\Omega \frac{1}{\mu} \left(\frac{\partial N_i}{\partial x} \frac{\partial N_j}{\partial x} + \frac{\partial N_i}{\partial y} \frac{\partial N_j}{\partial y} \right) d\Omega \qquad K_{pq} = - \int_\Omega \frac{1}{\mu} \frac{\partial N_i}{\partial p} \frac{\partial N_j}{\partial q} d\Omega$$

$$b_p = \int_\Omega N_{ip} J_p d\Omega \qquad\qquad\qquad (7.19)$$

where p and q refer to x, y or z. The overall matrix equation is similar to the previous ones, in that it is sparse. It is straightforward to show that the variational approach gives exactly the same result. Note that the gauge condition has not been used and is not required if **B** is the parameter of interest, which is usually the case in charged particle optics. The divergence condition on $\mathbf{A}$ is then not important. One way of imposing the divergence condition on $\mathbf{A}$ is to use edge elements (see chapter 4, section 7).

Another important feature about the above vector potential nodal approach is that it will yield and an asymmetric global matrix. This is evident from the entries in the cross-matrices, K_{pq}, which necessitates the use of more general iterative matrix inversion schemes than the ones normally used in finite element analysis.

2.   THE MAGNETIC SCALAR POTENTIAL IN THREE DIMENSIONS

2.1   The weighted residual approach

Consider Poisson's equation

$$\nabla \cdot \mu \nabla u = Q \tag{7.20}$$

where Q denotes a general source term, and μ is a scalar function which is dependent on spatial coordinates. Let the boundary conditions be expressed as

$$\mu(\partial u/\partial n) - p = 0 \quad \text{at } S_1 \quad \text{and} \quad u - u_0 = 0 \quad \text{at } S_2 \tag{7.21}$$

Let n be the number of unknown nodes in the domain, where the ith node is surrounded by m elements. Applying the weighted residual method at node i for Poisson's equation (see equations 4.108 and 4.109), we obtain

$$-\int_\Omega \mu \nabla N_i \cdot \nabla u_i d\Omega - \int_\Omega N_i Q d\Omega + \int_{S_1} N_i p dS = 0 \tag{7.22}$$

N_i in this equation is formed over contributions from m elements. The above equation can be represented by

$$[\mathbf{K}][\mathbf{u}] + [\mathbf{C}] = 0$$

where $[\mathbf{K}]$ is a n by n matrix, and $[\mathbf{u}]$ and $[\mathbf{C}]$ are 1 by n column vectors. The entries for $[\mathbf{K}]$ and $[\mathbf{C}]$ are given by

$$K_{ij} = -\int_\Omega \mu \nabla N_i \cdot \nabla N_j d\Omega$$

$$C_i = -\int_\Omega N_i Q d\Omega + \int_{S_1} N_i p dS \tag{7.23}$$

2.2   The Reduced Scalar Potential Formulation

Consider the divergence equation for the flux density in terms of the reduced scalar potential

$$\nabla \cdot \left(\mu \nabla \phi \right) = \nabla \cdot \left(\mu \mathbf{H}_s \right) = Q(x, y, z) \qquad (7.24)$$

For simplicity it will be assumed that the potential is either fixed on the outer boundary, or that its normal derivative is zero. Given that the reduced potential formulation is described by Poisson's equation it might seem that the matrices [$\mathbf{K}$] and [$\mathbf{C}$] can be simply written down in the following way

$$K_{ij} = -\int_{\Omega} \mu \nabla N_i \cdot \nabla N_j d\Omega \quad \text{and} \quad C_i = -\int_{\Omega} N_i \nabla \cdot \left(\mu \mathbf{H}_s \right) d\Omega \qquad (7.25)$$

But the source term is not trivial to evaluate in this form. Of course in the air region, it is effectively zero. For saturation, it has contributions from the iron region: the variation of $\mu(x,y,z)$ is directly generated by variations in the magnitude of $\mathbf{B}(x,y,z)$ which are known at each iteration of the non-linear calculation. Using the vector property that $\nabla.(f\mathbf{a})=f\nabla.\mathbf{a} + \mathbf{a}.\nabla f$, and that the divergence of $\mathbf{H}_s$ is zero, then the matrix [$\mathbf{C}$] at each iteration in the non-linear calculation

$$C_i = -\int_{\Omega_k} N_i \left(\mathbf{H}_s \cdot \nabla \mu \right) d\Omega_k \qquad (7.26)$$

More consideration however, needs to be given to the iron/air interface surface. Since there is an abrupt change in the permeability at this interface, any quantity that involves its spatial derivative is not simple to evaluate. One way around this problem is to split up the whole domain into iron regions Ω_k and air/source regions Ω_j (see Figure 1.10). Let the interface surfaces be represented by Γ_{jk}. A weighted residual can then be performed for each region separately. In the air/source region Ω_j, the residual equation is

$$R_{\Omega_j} = -\int_{\Omega_j} \mu_j \nabla N_i \cdot \nabla \phi d\Omega + \int N_i \mu_j \frac{\partial \phi}{\partial n} d\Gamma_{jk} \qquad (7.27)$$

and in the iron region Ω_k, for the non-saturated case

$$R_{\Omega_k} = -\int_{\Omega_k} \mu_k \nabla N_i \cdot \nabla \phi d\Omega - \int N_i \mu_k \frac{\partial \phi}{\partial n} d\Gamma_{jk} \qquad (7.28)$$

Note that the signs of the two surface integral terms are opposite to one another, this is because the direction normal to the interface depends upon which side it is approached. When combining these two equations, we obtain

$$R_{\Omega_j} + R_{\Omega_k} = -\int_{\Omega_j} \mu_j \nabla N_i \cdot \nabla \phi d\Omega - \int_{\Omega_k} \mu_k \nabla N_i \cdot \nabla \phi d\Omega + \int N_i \left(\mu_j \frac{\partial \phi_j}{\partial n} - \mu_k \frac{\partial \phi_k}{\partial n} \right) d\Gamma_{jk} = 0$$

$$(7.29)$$

A useful relation can be derived from noting that the normal component of the magnetic flux density across the interface is continuous (equation (1.86)), so

$$\mu_j \left(H_{sn} - \frac{\partial \phi}{\partial n} \right)_j = \mu_k \left(H_{sn} - \frac{\partial \phi}{\partial n} \right)_k \qquad (7.30)$$

where H_{sn} is the normal component of $\mathbf{H_s}$ at the interface. This equation can obviously be used to simplify the residual equation to

$$-\int_{\Omega_j} \mu_j \nabla N_i \cdot \nabla \phi d\Omega - \int_{\Omega_k} \mu_k \nabla N_i \cdot \nabla \phi d\Omega + \int N_i \left(\mu_j - \mu_k \right) H_{sn} d\Gamma_{jk} = 0 \qquad (7.31)$$

In general, **[K]** and **[C]** are given by

$$K_{ij} = -\int_{\Omega_j} \mu_j \nabla N_i \cdot \nabla N_j d\Omega_j - \int_{\Omega_k} \mu_k \nabla N_i \cdot \nabla N_j d\Omega_k$$

$$(7.32)$$

$$C_i = -\int_{\Omega_k} N_i \left(\mathbf{H}_s \cdot \nabla \mu \right) d\Omega_k + \int N_i \left(\mu_j - \mu_k \right) H_{sn} d\Gamma_{jk}$$

2.3   The two scalar potential formulation

The two scalar potential differential formulation is restated here for clarity. In the source/air region Ω_j have

$$\nabla \cdot \mu_j \nabla \phi = \nabla \cdot \mu_j \mathbf{H_s} \qquad (7.33)$$

In the iron region Ω_k

$$\nabla \cdot \mu_k \nabla \psi = 0 \qquad (7.34)$$

Since the intrinsic permeability will appear in all equations, only the relative permeability need be considered. Let μ_r be the relative permeability of the iron. In the source/air region of course, the relative permeability is unity. At the region interface, the boundary conditions (equations (7.30) and (1.88)) give

$$-\mu_r \frac{\partial \psi}{\partial n} = H_{sn} - \frac{\partial \phi}{\partial n} \qquad \text{and} \qquad -\frac{\partial \psi}{\partial t} = H_{st} - \frac{\partial \phi}{\partial t} \qquad (7.35)$$

In the region Ω_j,

$$R_{\Omega_j} = -\int_{\Omega_j} \nabla N_i \cdot \nabla \phi \, d\Omega + \int N_i \frac{\partial \phi}{\partial n} d\Gamma_{jk} \qquad (7.36)$$

In the region Ω_k, we have

$$R_{\Omega_k} = -\int_{\Omega_k} \mu_r \nabla N_i \cdot \nabla \psi \, d\Omega - \int N_i \mu_r \frac{\partial \psi}{\partial n} d\Gamma_{jk} \qquad (7.37)$$

Note that the boundary terms in the two residual expressions are of opposite sign, this comes from the fact that the sign of the boundary depends on which side it is approached. We can add the two residual equations together, giving

$$-\int_{\Omega_j} \nabla N_i \cdot \nabla \phi \, d\Omega - \int_{\Omega_k} \mu_r \nabla N_i \cdot \nabla \psi \, d\Omega + \int N_i \left(\frac{\partial \phi}{\partial n} - \mu_r \frac{\partial \psi}{\partial n} \right) d\Gamma_{jk} = 0 \qquad (7.38)$$

Substituting for the normal component at the interface yields

$$-\int_{\Omega_j} \nabla N_i \cdot \nabla \phi \, d\Omega - \int_{\Omega_k} \mu_r \nabla N_i \cdot \nabla \psi \, d\Omega + \int N_i H_{sn} d\Gamma_{jk} = 0 \qquad (7.39)$$

This residual equation gives the following type of matrix equation

$$[K]_{\Omega_j}[\phi] + [K]_{\Omega_k}[\psi] = [C]$$

where the entries of these matrices are

$$K_{ij} = \int_{\Omega} \mu_r \nabla N_i \cdot \nabla N_j \, d\Omega \qquad C_i = \int_{\Gamma} N_i H_{sn} d\Gamma_{jk} \qquad (7.40)$$

The above matrix equation is valid for interior nodes, but for nodes on the interface there is ambiguity as to which potential to use. Continuity of the tangential H-field along a path A to B on the interface gives the following relationship between the potentials

$$\phi_A - \phi_B = \psi_A - \psi_B + \int_A^B H_{st} d\Gamma \qquad (7.41)$$

Let the two potentials be equal at the point A. So when in the iron region, it is possible to eliminate ϕ_j on the interface by

$$\sum_j \nabla N_i \cdot \nabla N_j \phi_j = \sum_j \nabla N_i \cdot \nabla N_j \left(\psi_j - \int_A^{B_j} H_{st} d\Gamma \right)$$ (7.42)

where B_j is the jth node along the interface. This generates an extra term on the right hand side, so that the matrix coefficients for the iron region are given by

$$K_{ij} = \int_{\Omega_k} \mu_r \left(\nabla N_i \cdot \nabla N_j \right) d\Omega_k \qquad C_i = \int N_i H_{sn} d\Gamma_{jk} + \sum_j \nabla N_i \cdot \nabla N_j \left(\int_A^{B_j} H_{st} d\Gamma \right)$$

(7.43)

The total potential in the iron region can now be calculated. For the region Ω_j, a similar procedure is used, ψ_j is eliminated along the interface, giving an explicit boundary condition on the reduced scalar potential.

3.   SATURATION EFFECTS

For non-linear problems, the Newton Raphson method can be used. Let the potential distribution have values on n nodes, u_1.......u_n. At the ith node, the residual r_i is dependent on the nodal potentials

$$r_i = f_i(u_1, u_2 ....u_n)$$ (7.44)

In practice, f_i is a function which depends on the potential at neighbouring nodes, and does not involve all the nodes in the domain.

These non-linear systems can be solved in the following iterative way. If the nodal potential distribution at the kth iteration is denoted by $[\mathbf{u}]^{(k)}$, a column vector of rank 1 by n, and $[\mathbf{r}]^{(k)}$ is its corresponding residual column vector, then the Newton-Raphson iteration scheme proceeds to calculate an improved estimate of the potential $[\mathbf{u}]^{(k+1)}$ by

$$[\mathbf{u}]^{(k+1)} = [\mathbf{u}]^{(k)} - [\mathbf{J}]^{-1}[\mathbf{r}]^{(k)} \qquad \text{where} \qquad J_{ij} = \frac{\partial r_i}{\partial u_j}$$ (7.45)

The change of the nodal potential values at each iteration $[\Delta\mathbf{u}]^{(k)}$ can be expressed in the form of n linear set of equations

$$[\mathbf{J}][\Delta\mathbf{u}]^{(k+1)} = [\mathbf{r}]$$ (7.46)

It turns out that this matrix problem has an identical structure to the one used to solve the linear potential solution, so no additional matrix solver is required. Note that at every iteration of the non-linear calculation, the matrix $[\mathbf{J}]$ needs to be updated, and the above matrix equation needs to be solved. Since the entries J_{ij} change at each iteration, it is not possible to store them and pre-factorise a single matrix. The assembly and solution of a new matrix equation at each iteration means the non-linear solution will take much longer to execute than the solution for linear problems. Fortunately, the Newton-Raphson algorithm has a fast convergence property, and in many cases the number of iterations can be kept below ten.

Recall that the ith row of the matrix equation for linear systems has a residual equation of the following form

$$\left[\sum_j K_{ij} u_j\right]_{A_{e1}} + \left[\sum_j K_{ij} u_j\right]_{A_{e2}} + \ldots \ldots \ldots \left[\sum_j K_{ij} u_j\right]_{A_{em}} - b_i = r_i = 0 \qquad (7.47)$$

K_{ij} are the stiffness matrix entries and b_i comes from the boundary conditions and source distribution (equation 7.6). Now it is obvious that J_{ij} is then given by

$$J_{ij} = \frac{\partial r_i}{\partial u_j} = \frac{\partial}{\partial u_j}\left(\left[\sum_j K_{ij} u_j\right]_{\Omega e_1} + \ldots \left[\sum_j K_{ij} u_j\right]_{\Omega e_m}\right) \qquad (7.48)$$

This is the general expression and it is valid for all magnetic problems.

Consider the two dimensional form of the vector potential A(r,z) in axi-symmetric coordinates, where u(r,z)=rA. In this case

$$K_{ij} = 2\pi\iint_{\Omega_e} \frac{1}{\mu r}\left[\left(\frac{\partial N_i}{\partial z}\right)\left(\frac{\partial N_j}{\partial z}\right) + \left(\frac{\partial N_i}{\partial r}\right)\left(\frac{\partial N_j}{\partial r}\right)\right]drdz \qquad (7.49)$$

It is useful here to express K_{ij} in terms of the reluctivity, $\upsilon=\mu^{-1}$ and a function p(r,z) defined in terms of the flux density **B** by $p=|\mathbf{B}|^2$, so

$$p = \frac{1}{r^2}\left[\left(\frac{\partial u}{\partial r}\right)^2 + \left(\frac{\partial u}{\partial z}\right)^2\right] \qquad (7.50)$$

By expanding p in terms of the potential u(r,z), and taking its derivative with respect to the potential at the ith node, it is simple to show that

$$r\frac{\partial p}{\partial u_i} = \frac{2}{r}\sum_j\left(\frac{\partial N_i}{\partial r}\frac{\partial N_j}{\partial r} + \frac{\partial N_i}{\partial z}\frac{\partial N_j}{\partial z}\right)u_j \qquad (7.51)$$

Note also that

$$r \frac{\partial p}{\partial u_j \partial u_i} = \frac{2}{r} \left(\frac{\partial N_i}{\partial r} \frac{\partial N_j}{\partial r} + \frac{\partial N_i}{\partial z} \frac{\partial N_j}{\partial z} \right) \tag{7.52}$$

K_{ij} can then be related to the function p by

$$\sum_j K_{ij} u_j = \pi \iint_{\Omega_e} r \upsilon \frac{\partial p}{\partial u_i} dr dz \tag{7.53}$$

To find J_{ij}, this expression needs to be differentiated with respect to u_j. It is convenient to take the reluctivity to be a function of p, $\upsilon(p)$, which can be found from the B-H curve of the magnetic material, so

$$\frac{\partial \upsilon}{\partial u_j} = \frac{\partial \upsilon}{\partial p} \frac{\partial p}{\partial u_j} \tag{7.54}$$

The entries for [J] are then given by

$$J_{ij} = \pi \iint_{\Omega_e} r \frac{\partial \upsilon}{\partial p} \frac{\partial p}{\partial u_i} \frac{\partial p}{u_j} dr dz + K_{ij} \tag{7.55}$$

Note that p and its derivatives with respect to u_i and u_j depend on the potential distribution [u] and can easily be calculated at each iteration in the Newton-Raphson scheme.

The same method can be applied to the scalar potential $\phi(r,z)$, where $\mathbf{H}=-\nabla\phi$. Let the function $q=|H|^2$. In three dimensional cartesian coordinates,

$$K_{ij} = \int_{\Omega} \left[\left(\frac{\partial N_i}{\partial x} \right) \left(\frac{\partial N_j}{\partial x} \right) + \left(\frac{\partial N_i}{\partial y} \right) \left(\frac{\partial N_j}{\partial y} \right) + \left(\frac{\partial N_i}{\partial z} \right) \left(\frac{\partial N_j}{\partial z} \right) \right] d\Omega \tag{7.56}$$

So the residual r_i in this case is given by

$$r_i = \sum_j K_{ij} u_j = \int_{\Omega} \left[\left(\frac{\partial N_i}{\partial x} \right) \left(\frac{\partial u}{\partial x} \right) + \left(\frac{\partial N_i}{\partial y} \right) \left(\frac{\partial u}{\partial y} \right) + \left(\frac{\partial N_i}{\partial z} \right) \left(\frac{\partial u}{\partial z} \right) \right] d\Omega \tag{7.57}$$

Here the derivatives of u(x,y,z) are related to their corresponding shape function expansions. The function q and its derivatives with respect to the potential at nodes i and j are obviously given by

$$q = \left(\frac{\partial u}{\partial x}\right)^2 + \left(\frac{\partial u}{\partial y}\right)^2 + \left(\frac{\partial u}{\partial z}\right)^2$$

$$\frac{\partial q}{\partial u_i} = 2\left[\left(\frac{\partial N_i}{\partial x}\right)\left(\frac{\partial u}{\partial x}\right) + \left(\frac{\partial N_i}{\partial y}\right)\left(\frac{\partial u}{\partial y}\right) + \left(\frac{\partial N_i}{\partial z}\right)\left(\frac{\partial u}{\partial z}\right)\right]$$ (7.58)

$$\frac{\partial^2 q}{\partial u_j \partial u_i} = 2\left[\left(\frac{\partial N_i}{\partial x}\right)\left(\frac{\partial N_j}{\partial x}\right) + \left(\frac{\partial N_i}{\partial y}\right)\left(\frac{\partial N_j}{\partial y}\right) + \left(\frac{\partial N_i}{\partial z}\right)\left(\frac{\partial N_j}{\partial z}\right)\right]$$

The residual expressed in terms of q is

$$r_i = \int_\Omega \frac{\mu}{2}\frac{\partial q}{\partial u_i}d\Omega$$ (7.59)

Then the entries of [**J**] for each element are found from

$$J_{ij} = K_{ij} + \frac{1}{2}\int\frac{\partial \mu}{\partial q}\frac{\partial q}{\partial \phi_i}\frac{\partial q}{\partial \phi_j}d\Omega$$ (7.60)

The function q and its derivatives with respect to the potential nodes are known at each iteration of the Newton-Raphson scheme, so the above expression can easily be calculated. For axisymmetry it follows that

$$J_{ij} = K_{ij} + \pi\iint_{\Omega_e} r\frac{\partial \mu}{\partial q}\frac{\partial q}{\partial \phi_i}\frac{\partial q}{\partial \phi_j}drdz$$ (7.61)

The foregoing method formulates the problem in terms of the residual vector [**r**]. A variational can also be used, but it is more complicated. In this case the functional F for the vector potential is

$$F = \int_\Omega (W - J\cdot A)$$ (7.62)

W represents the energy density at each point and is given by

$$W = \int_0^B H\cdot dB = \int_0^B \frac{b}{\mu}db$$ (7.63)

where b is a dummy variable. This energy density W can be expressed in terms of the reluctivity υ and the function p, where p=B²

$$W = \frac{1}{2}\int_0^{B^2} \upsilon dp$$ (7.64)

The derivative of W with respect to u_i is then given by

$$\frac{\partial W}{\partial u_i} = \frac{\partial W}{\partial p}\frac{\partial p}{\partial u_i} = \frac{1}{2}\upsilon\frac{\partial p}{\partial u_i} \qquad (7.65)$$

Note that this differentiated again by u_j gives

$$\frac{\partial^2 W}{\partial u_j \partial u_i} = \frac{1}{2}\frac{\partial \upsilon}{\partial p}\frac{\partial p}{\partial u_j}\frac{\partial p}{\partial u_i} + \frac{1}{2}\upsilon\frac{\partial^2 p}{\partial u_j \partial u_i} \qquad (7.66)$$

Now minimizing F with respect to the potential at the ith node, taking into account contributions from surrounding elements, gives the residual r_i

$$r_i = \frac{\partial F}{\partial u_i} = \int_\Omega \left(\frac{\partial W}{\partial u_i} - JN_i\right)d\Omega \qquad (7.67)$$

Here it is assumed that there is only one component of the current density $\mathbf{J}$. The entries of the Jacobian matrix J_{ij} are then given by

$$J_{ij} = \frac{\partial^2 F}{\partial u_j \partial u_i} = \int_\Omega \left(\frac{\partial^2 W}{\partial u_j \partial u_i}\right)d\Omega \qquad (7.68)$$

which is simplified to

$$J_{ij} = \int_\Omega \left(\frac{1}{2}\frac{\partial \upsilon}{\partial p}\frac{\partial p}{\partial u_j}\frac{\partial p}{\partial u_i}\right)d\Omega + \int_\Omega \left(\frac{1}{2}\upsilon\frac{\partial^2 p}{\partial u_j \partial u_i}\right)d\Omega \qquad (7.69)$$

This expression can be formulated for different coordinate systems. It is quite simple to show that it is identical to the one already derived for the vector potential in axisymmetric coordinates.

For the scalar potential, the starting functional is given by

$$F = \int_\Omega (W)d\Omega \qquad (7.70)$$

and W is given by

$$W = \int_0^B \mathbf{H}\cdot d\mathbf{B} = \int_0^H \mu h\,dh \qquad (7.71)$$

The same procedure is used as before, expressing the result in terms of the function q, where $q=|H|^2$.

Chapter 8
ELECTRIC LENSES

1. ACCURACY ISSUES

1.1 The two-tube lens example

To estimate the accuracy of the finite element method for electric lenses, different ways of implementing it need to be critically compared together on test structures which have analytical field solutions. The test example chosen here is the well-known two-tube cylindrical lens whose layout is depicted in Figure 8.1a.

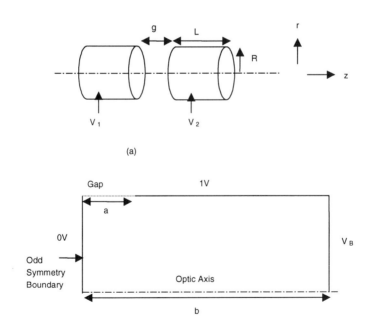

(a)

Figure 8.1 The two-tube test lens example
(a) Schematic layout   (b) Simulation model

Two metal tubes of length L are separated by a gap g and biased at the voltages V_1 and V_2. The potential distribution of the potential distribution inside the lens $V(r,z)$ can be expressed in terms of the field distribution $V_N(r,z)$ of a simpler layout which uses an odd symmetry plane at the centre of the lens (z=0), as shown in Figure 8.1b. Here it is easy to show that

$$V(r,z) = \frac{1}{2}(V_1 + V_2) + \frac{1}{2}(V_2 - V_1)V_N(r,z) \qquad (8.1)$$

Linear interpolation is used in the gap between the plane of odd symmetry and the 1 volt conductor, this is a closed boundary approximation to the real two-tube lens in which the gap is an open boundary. The thickness of the plates are neglected in the numerical model and represented by equipotential planes. The boundary at the end of the tube, denoted by V_B in Figure 8.1b, is specified explicitly to be 1 volt. The tube's length was made much greater than its radius, so that the boundary condition at the end of the tube does not greatly effect the potential distribution inside the lens. In some situations, in order to reduce computation time and memory requirements, an artificial boundary was specified at a shorter distance z from the centre of the lens. The potential distribution $V_B(r)$ on this boundary was derived from the analytical field solution.

The radius R of the electrodes is taken to be 5 mm and the electrode gap g is 1 mm. The total length of the lens typically ranges from z=-30 mm to z=30 mm. At this diameter to length ratio, the boundary condition at the ends of the tubes do not significantly affect its optical properties. The voltages V_1 and V_2 are specified to be 1000 and 10000 volts respectively.

Let a and b denote distances from the plane of odd symmetry to the beginning and ends of the 1 volt tube respectively, as shown in Figure 8.1b. The analytical formula for the field distribution inside the lens can be found by using a Fourier-Bessel series expansion. $V_N(r,z)$ is given by the following expression [1]

$$V_N(r,z) = 2\sum_{m=0}^{\infty} \frac{\sin(k_m a)}{k_m a} \frac{\sin(k_m z)}{k_m b} \frac{I_0(k_m r)}{I_0(k_m R)} \quad \text{where} \quad k_m = \frac{\pi}{b}\left(m + \frac{1}{2}\right) \qquad (8.2)$$

A thousand terms in the above series gave better than 6 digit accuracy on the potential distribution close to the optic axis. The electric field components at any point can obviously be derived from the above formula.

1.2   First and second-order elements

Before discussing the results, it is instructive to examine the Taylor's series expansion for the potential close to the axis. As already mentioned in Chapter 2, section 2, the potential $V(r,z)$ can be expressed in terms of a series expansion in r and the axial potential distribution $V_0(z)$

$$V(r,z) = V_0(z) - \frac{r^2}{4}\frac{\partial^2 V_0(z)}{\partial z^2} + \frac{r^4}{64}\frac{\partial^4 V_0(z)}{\partial z^4} - \frac{r^6}{2304}\frac{\partial^6 V_0(z)}{\partial z^6} + \dots\dots \quad (8.3)$$

From this equation, it is possible to obtain the Taylor's expansion for $V(r,z+\Delta z)$

$$V(r,z+\Delta z) = V_0 + \Delta z\frac{\partial V_0}{\partial z} + \left(\frac{\Delta z^2}{2} - \frac{r^2}{4}\right)\left(\frac{\partial^2 V_0}{\partial z^2}\right) + \left(\frac{\Delta z^3}{6} - \frac{\Delta z r^2}{4}\right)\left(\frac{\partial^3 V_0}{\partial z^3}\right)$$
$$+ \left(\frac{\Delta z^4}{24} - \frac{\Delta z^2 r^2}{8} + \frac{r^4}{64}\right)\left(\frac{\partial^4 V_0}{\partial z^4}\right)$$

$$(8.4)$$

If the mesh spacing is given by h in the r direction, it is clear from this expression that the errors for a first-order approximation of the axial potential will typically be proportional to h^2, Δz^2 and the double derivative of $V_0(z)$. For a second-order approximation, obviously the error will be generated by higher terms. Exactly which terms are present depends on the form of the governing equation and the numerical approximation to it. If for instance, the nodes in both r and z are equidistant from the centre node, then $V(r,z+\Delta z) + V(r,z-\Delta z)$ in the governing equation will obviously generate a truncation error which is primarily proportional to h^4 and the forth derivative of the axial potential.

The central sections of three mesh layouts are depicted in Figures 8.2a-c. The first utilizes a regularly spaced mesh, where mesh lines are separated by 0.25 mm. The second layout is a block region mesh, where the mesh spacing is constant within a region, but varies from region to region. The central region in Figure 8.2b spans an area of 0.05 mm in r and 0.5 mm in z and uses 11 by 11 mesh lines. Mesh region boundaries are located at r=0.05, 0.15, 0.35, 0.75, 1.55, 3.15, 5 mm, and z=0.5, 1.5, 3.5, 7.5, 15.5, 30 mm. The mesh spacing is doubled at a region boundary. The third type of mesh is a graded mesh, similar to the one reported by Lencová [2] and is one where the mesh spacing increases gradually in both r and z directions. Here 21 by 21mesh lines are evenly spaced over a central region spanning 0.5mm by 0.5 mm, beyond which the mesh spacing is progressively expanded by a factor of 1.05 from mesh cell to mesh cell in both r and z directions. In the following study, the mesh resolution for all mesh types is systematically varied, and the maximum error on the axial potential is calculated.

For the first-order element results in this section, triangles are used which incorporate the r^2 correction. A 3pt integration is employed to calculate the local stiffness matrix terms. For the second-order element results, 9 node quadrilateral elements are used with 6pt by 6pt Guassian Quadrature integration.

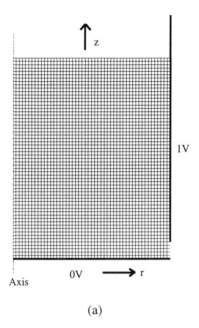

(a)

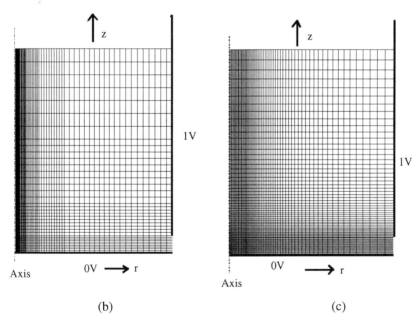

(b)   (c)

Figure 8.2 Numerical Mesh Layouts
(a) Regular
(b) Region block
(c) Graded

Figure 8.3 shows the potential and its first few derivatives as predicted from the analytical formula. This graph is helpful in relating the axial error distribution to specific terms in the Taylor's series expansion of the potential.

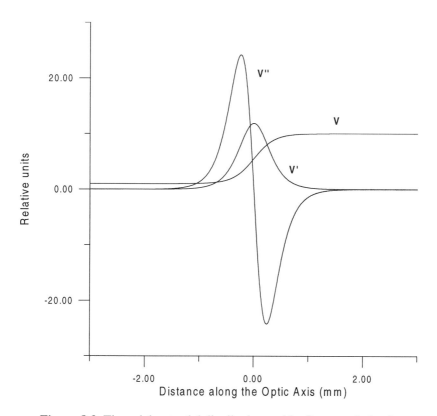

Figure 8.3 The axial potential distribution and its first two derivatives

The first results are generated for first-order triangle elements and are shown in Figure 8.4a. They confirm that the error on the axial potential varies as h^2 and that it is approximately proportional to the second derivative of the axial potential distribution. These results were produced from the graded mesh layout. The axial electric field strength is found by using cubic spline interpolation on the axial potential distribution. The spline interpolation procedure faithfully reproduces the h^2 dependence and does not significantly add to the error. The form of the error on the electric field, shown in Figure 8.4b, is as expected, proportional to the third derivative of the axial potential distribution. Similar results in Figures 8.5a and 8.5b are shown for the region block mesh type. The discontinuities in the graph illustrate the effect of abrupt mesh spacing changes at region boundaries. These discontinuities clearly increase the axial electric field error.

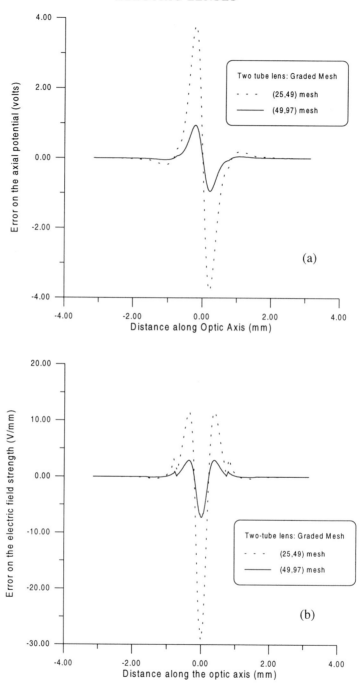

Figure 8.4 Axial errors on the graded mesh layout
(a) Potential   (b) Field

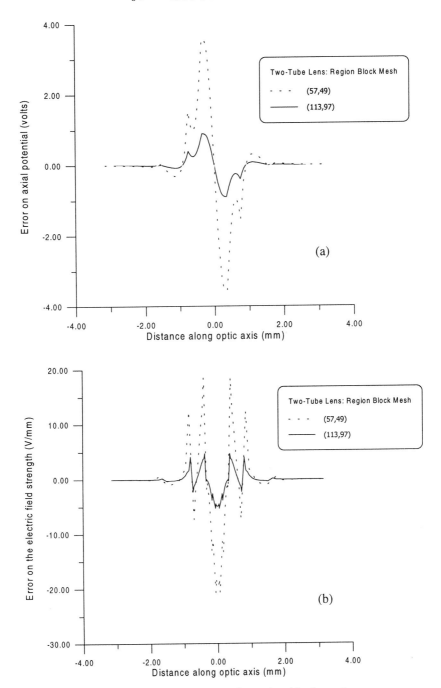

Figure 8.5 Axial errors on the region block mesh
(a) Potential   (b) Field

The error for 9 node second-order elements was found to be approximately proportional to h^4. This variation is illustrated in Figure 8.6 for the graded mesh layout. The form of the derivative appears to be proportional to the forth derivative of the axial potential distribution. Although the program run times for the second-order element solution were typically between 10 to 15 times longer than those for first-order elements on the same mesh, their corresponding errors were around 100 times smaller. This result indicates that second-order elements provide greater accuracy than first-order elements. On the other hand, Richardson's extrapolation works better on first-order element solutions than on second order elements. The h^2 error variation for first-order elements was found to be more predictable than the h^4 error variation for second-order elements. Figure 8.7 shows an example where the second-order element errors of the extrapolated distribution from the mesh resolutions (25,49) and (49,97) on the graded mesh layout are actually worse than the errors produced by the initial (49,97) mesh. From the mesh resolutions (13,25) and (25,49), the extrapolated results exhibited a factor of three improvement. On the other hand, the factor of improvement for first-order elements after extrapolation typically varied between 10 to 100. Figure 8.8 shows that the extrapolation errors from the (13,25) and (25,49) mesh layouts using first-order elements were only marginally worse than the second-order solution using the (25,49) mesh, despite them being 12 times faster to run. Figure 8.9 summarises these results for the regular mesh. Similar results were obtained for the graded and region block meshes.

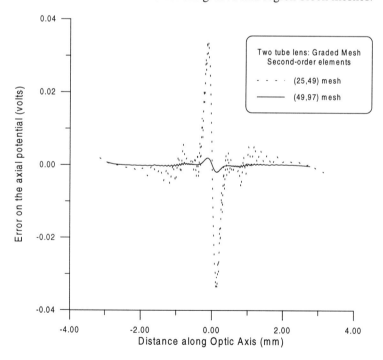

Figure 8.6 Axial potential error for second-order finite elements

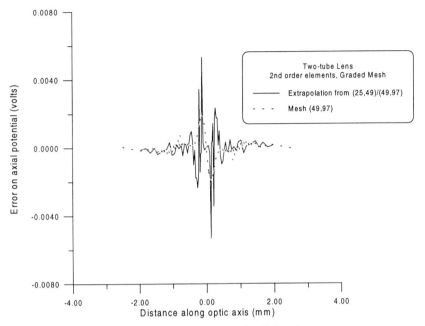

Figure 8.7 Extrapolation on second-order finite elements

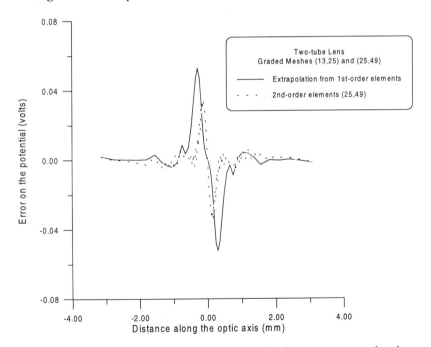

Figure 8.8 Comparison of extrapolated first-order elements to second-order element solution

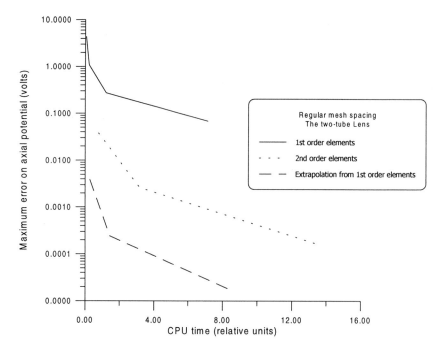

Figure 8.9 Comparison of first-order with second-order element accuracy as a function of computation time

Figures 8.10a and 8.10b show how the accuracy of the region block mesh compares with the graded one. For both first-order and second-order elements, clearly a graded mesh is superior to a region block mesh.

To summarise the results for first and second order elements: in a given amount of program run time, although second-order elements are more accurate than first-order elements, Richardson's extrapolation works better for first-order elements than second-order elements, thus giving an overall advantage to first-order elements. Also, mesh layout is important to the final accuracy of the axial field distribution. Where possible, graded meshes should be used in preference to region block meshes.

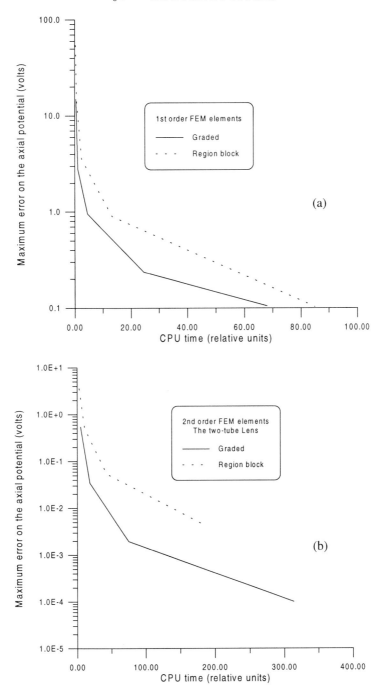

Figure 8.10 Comparison of graded and region block meshes
(a) First-order elements   (b) Second-order elements

1.3   Cubic elements

To investigate the accuracy of cubic elements, it is convenient to use serendipity elements. A single finite element program using such elements can be programmed to vary the order of the element as required. The nodes of the original mesh become element corner nodes, and the element order is increased by adding extra nodes to element edges. This procedure provides the flexibility of specifying mixed-order elements. The results are presented in Figures 8.11a and 8.11b. Figure 8.11a presents the axial potential error for first, second and third order elements as a function of the number of nodes used. As expected, third-order elements are the most accurate. For a given mesh, cubic elements are the best choice. But this is not a fair comparison since the number of non-zeros in the final global matrix increases as the element order rises: for higher order elements, more surrounding nodes are involved in the nodal equation. Figure 8.11b shows how the axial potential error varies with computation time, which is a more meaningful comparison. In this case, second-order elements are more accurate than third-order elements. The reason for the better performance of second-order elements is due to how the potential varies as a function of radius. From the Taylor's series given by equation (8.3), the potential varies in even powers of r. This means that a second-order approximation is better suited to model the radial variation of the potential than a third-order one: both second-order and third-order in the radial direction have truncation errors which come from potential variations which are proportional to r^4, and for this reason, second-order elements perform better than third-order elements.

The axial potential error for third-order elements did not fall in a predictable way as the mesh resolution was systematically doubled, so extrapolation did not significantly improve the results. If extrapolation is used, for a given amount of computation time, first-order elements give the best accuracy performance.

Mixed-order elements can be used around the optic axis with serendipity elements. This form of high-order refinement is expected to improve the accuracy for a given amount of computation time. Second-order elements can for instance be used around the optic axis of a first-order element mesh layout. This procedure not only improves accuracy but also provides the necessary r^2 correction of the potential around the optic axis.

It should be noted that the first-order element results shown in Figure 8.11 are different to the ones presented in the previous section. Here 6pt by 6pt high-order Gaussian Quadrature integration is used on quadrilateral serendipity elements, whereas in the previous section, low-order integration (typically 3pt) is used on triangle elements.

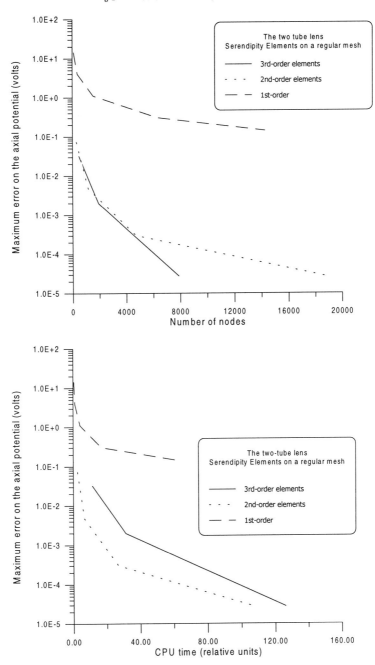

Figure 8.11 Axial potential error on cubic elements
(a) As a function of mesh nodes   (b) As a function of computation time

2.   DIRECT RAY TRACING USING OFF-AXIS MESH NODE POTENTIALS

2.1   Direct ray tracing vs perturbation methods

Perturbation of the paraxial trajectory has disadvantages when the primary electron beam traverses regions far from the optic axis since they require the use of high-order derivatives of the axial potential. Using high-order axial derivatives can place severe constraints on the continuity of the axial field distribution. The Taylor's series expansions of the potential must include high enough terms to calculate the aberration coefficients of interest. To calculate the 5th order aberrations of electric lenses for instance, the fourth derivative of the numerically solved potential distribution along the optic axis is required [3]. To calculate 5th order aberrations of magnetic lenses, it is the third derivative of the axial flux density which is required. In addition, the calculation of high-order aberration coefficients usually requires the derivation of complicated integration formulas which are time consuming to check and program. To minimize errors incurred from interpolation, quintic spline functions can be used along the optic axis. But this still leaves truncation errors that come from the field solution, as well as trajectory integration errors.

A direct ray trace technique has the merit of being much simpler than the conventional aberration integral method, but it requires high accuracy calculation of the field values along the trajectory path. By plotting trajectories for a variety of different initial conditions at the object plane and registering their positions and slopes on the image plane, both the final probe size and aberration coefficients can be calculated.

There are several ways to plot charged particle trajectories directly. The simplest approach is to solve the equations of motion for the charged particle in time, that is, solve the standard Lorentz equation. Given the electric field $E(r,z)$ everywhere in axisymmetric cylindrical coordinates (r,z) and the particle's initial position and velocity components in Cartesian coordinates $x(0)$, $y(0)$, $z(0)$, $v_x(0)$, $v_y0)$, and $v_x(0)$ respectively, then there are well-known integration methods such as Runge-Kutta and Predictor-Corrector methods which can be used to solve for $x(t)$, $y(t)$, $z(t)$. The position and slope of each ray at the image plane is monitored.

Accurate field values close to the axis calculated from the potential at off-axis mesh nodes is not a straightforward task, and necessarily involves issues of mesh topology. The mesh density needs to be relatively high around the axis and high-order field interpolation is required. This is not a trivial task when using the finite element method since mesh lines are often unstructured and do not generally intersect at right angles. Mesh truncation errors as well as trajectory integration errors are also important.

The following study investigates the precision to which the third-order spherical aberration coefficient C_{s3} and fifth order spherical aberration coefficient C_{s5} can be derived by direct ray tracing on the simple two-tube test electric lens layout shown in Figure 8.1. The change of the focal position Δz from the Gaussian plane in a lens limited purely by spherical aberration can be given in terms of the following power series

$$\Delta z = C_{s3}\alpha^2 + C_{s5}\alpha^4 + ...$$ (8.5)

where α is the final semi-angle. Trajectories are traced for parallel rays that start at different radial positions r_a. The point of focus for each ray z_f is noted and stored. The final semi-angle is calculated from the focal length f and input aperture radius ($\alpha = r_a/f$). By using a least squares fit on the above equation linking Δz and α, the values of C_{s3} and C_{s5} are estimated.

The first-order finite element formulation used here is similar to that given by Lencová [4], where the local element basis function varies linearly with r^2 and z. For second-order elements, a 9 node quadrilateral element is used, similar to that employed by Munro [5]. All field solving programs used were written by the author, some of which can be found in the KEOS package [6].

For the plotting of electron trajectories, a mesh field interpolation method based upon bi-cubic splines is used. This method for non-rectangular mesh cells was first reported by Khursheed, and consists of numerically transforming quadrilateral regions in real space on to square regions in normalised space. Bi-cubic spline interpolation is used in normalised space to find normalised field values, which are then transformed back into electric field components in real space. The method was reported to provide a fast and accurate way by which electron trajectories can be plot on finite element solved field distributions [7] (see also chapter 11). In this case, since all mesh cells are either square or rectangular in shape, bi-cubic spline can be applied directly to the mesh in real space and provides a good comparison with the more general method developed by Khursheed. The first-order derivatives and cross-derivatives are pre-computed at every mesh node by a series of cubic splines.

To integrate the equations of motion, a variable-step Runge-Kutta Merson method is used [8] The method is able to vary the trajectory length according to its own estimate of the relative trajectory step truncation error.

When using direct ray tracing, a C_{s3} value is extrapolated from three sets of twenty trajectories, each set having initial radii less than 0.1R, 0.02R and 0.01R respectively (R is the two-tube lens radius). For perturbation of the paraxial trajectory, the analytical Fourier-Bessel formula is used to generate 1200 on-axis potential data points and is input into a KEOS program (ABAXIS) which calculates on-axis aberrations coefficients. The dimensions of the test two-tube lens is similar to the previous section, where the lens gap size is one tenth of the tube diameter and the voltage on the right-hand side tube is ten times that of the voltage of the left-hand side tube. The tube length is assumed to be infinite. In keeping with the convention used for previously published data on the two-tube lens, all optical parameters are normalised to the tube diameter, and are therefore dimensionless.

2.2   Trajectory integration errors

An agreement of better than four digits is obtained for the focal points and focal lengths derived from plotting the paraxial ray. Rays travelling in the positive z-direction focus at the point F_2, which by direct ray tracing is calculated to be 1.177589. From tracing the paraxial ray, it is 1.177588, giving an agreement of around 0.000085%. The focal length for rays travelling in the negative z-direction f_1 is found to be 0.79887 through direct ray tracing, while from the paraxial ray it is found to be 0.79882, giving an agreement of around 0.0062%. These results show that an agreement of typically better than 0.01% is obtained for the first-order properties of the lens.

When direct ray tracing is used to determine C_{s3}, a value of 10.0312 is obtained. The corresponding value found from 3rd-order perturbation of the paraxial trajectory is 10.0269, giving an agreement to within 0.042%. These results demonstrate the high accuracy to which the direct ray trace method can derive on-axis aberration coefficients, if the field distribution is known analytically.

The above value of C_{s3} derived from direct ray tracing can be compared with previous work. Renau and Heddle for instance, used a variational method to describe the potential and a ray tracing procedure involving three spatial derivatives of the potential [9]. They calculated a value of 12.6 for the aberration coefficient m_{26}, which is related to C_{s3} by $C_{s3}=f_1m_{26}$. Using the value of f_1 calculated in the present work, the value of C_{s3} by Renau and Heddle is 10.06511. This gives an agreement with the present value of C_{s3} (10.0312) to better than 0.034%.

To calculate the fifth-order spherical aberration coefficient C_{s5}, larger input radii were taken. The maximum input radius of the rays was increased to 60% of the lens-tube radius (R) and the number of trajectories was increased to 50. In the least squares fit, a three parameter fit was made, including the effect of the 7th order spherical aberration coefficient. The values obtained for C_{s3} were close to values obtained by a single parameter fit at a much smaller radius. The value of C_{s5} derived from the analytical field solution for these conditions was -106.390 for 500 trajectory steps. This value agrees well with results reported by Renau and Heddle. They calculated a value of -130 for the aberration coefficient q_{26}, which is related to C_{s5} by $C_{s5}=f_1 q_{26}$. Using the value of f_1 calculated in the present work, the value of C_{s5} given by Renau and Heddle is 103.85. This gives an agreement of around 2.5%. Since they quote an accuracy of 20% for C_{s5}, the value found from the present work seems reasonable. Moreover, it is not the precise value of C_{s5} that is important here, but its variation due to interpolation, truncation and trajectory integration errors.

By ray tracing directly from the analytical formula for the field distribution, trajectory integration errors are investigated. They are found to be dependent not only on the number of steps along the trajectory path, but also on how close the rays are to the optic axis.

Table 8.1 presents values of C_{s3} and shows how it varies with the number of trajectory steps in a ray and maximum input aperture radius.

C_{s3} values	Number of Steps in a Trajectory Path			
Maximum input aperture radius	90	159	281	500
0.1R	9.9903	9.983796	9.98456	9.98571
0.01R	10.245	9.84385	10.02603	10.03146
0.02R	9.777	10.0564	10.0143	10.0165
0.0 (extrapolated)	10.0353	9.9349	10.0268	10.0312

Table 8.1 Dependence of the third-order spherical aberration coefficient on trajectory integration steps and the maximum input aperture radius

Table 8.1 shows that variations in C_s due to trajectory integration errors are small. Just below 500 steps, they lie below 0.054%, while beyond 500 steps, they are typically less than 0.0025%. As expected, more precision is required for rays that are closer to the axis. The C_{s5} value varied from 107.161, 106.558, 106.39, and 106.587 as the number of steps in the trajectories was varied from 159, 282 and 500, and 900 respectively. Beyond 1000 steps, fifth order spherical aberration coefficient variations due to trajectory integration errors are kept below 0.1%.

2.3   Field interpolation errors

To investigate the effect of mesh field interpolation errors, rays were traced through the numerical meshes depicted in Figures 8.2b and 8.2c. The graded mesh shown in Figure 8.2c is designed to reduce the effect of truncation errors by having square mesh cells in the lens central area (0.1R by 0.1R), and gradually expanding the mesh spacing in both directions by a factor of 1.05. The block region mesh shown in Figure 8.2b uses a central area of 0.01R in the radial direction by 0.1R in the longitudinal direction, and is designed to minimize field interpolation errors in the radial direction. In the previous section, it was shown that the graded mesh gives a more accurate axial potential distribution than that produced by the region block mesh for the same program run-time. Where aberration coefficients are derived by the axial potential and its derivatives, then it is recommended that the graded mesh be used. On the other hand, it is obvious that where aberration coefficients are derived by direct ray tracing using the potential at off-axis mesh nodes, the relative performance between the two mesh layouts will be quite different. In this case, it is the mesh spacing in the radial direction that limits the accuracy to which aberration coefficients can be calculated, and the region block mesh shown in Figure 8.2b gives more accurate results than the graded mesh depicted in Figure 8.2c.

Tables 8.2a and 8.2b show results for the percentage accuracy of C_{s3} derived by near-axis direct ray tracing on the graded mesh and region block meshes and how it varies with the total number of nodes used. A first-order finite element solution is compared with the one that uses analytical values at the mesh nodes.

Number of nodes	% error on C_{s3} from analytical nodal potentials	% error on C_{s3} from first-order FEM solution
325	12.67	11.09
1225	0.197	0.1
4753	0.01	0.028
18,721	0.007	0.02

(a)

Number of nodes	% error on C_{s3} from analytical nodal potentials	% error on C_{s3} from first-order FEM solution
725	5.01	39.7
2,793	0.153	7.66
10,961	0.012	1.78
43,425	0.003	0.44

(b)

Table 8.2 Percentage accuracy of C_{s3}
(a) Graded mesh layout
(b) Region block mesh layout

For a low number of mesh nodes on the graded mesh, the accuracy between the finite element solution and analytical field solution is comparable, so the error is clearly limited by interpolation errors. As the number of mesh nodes increase, the error is clearly dominated by truncation noise. On the other hand, for the block region mesh, which has more mesh lines in the radial direction than in the longitudinal direction, the first-order finite element solution is much poorer in performance than the one derived from the analytical field solution for all mesh sizes, indicating that the error is dominated by truncation errors. These results indicate that when deriving C_{s3} by near-axis direct ray tracing, it is important to use small mesh spacing in the radial direction close to the optic axis.

Table 8.3 tabulates the errors for C_{s5} on the region block mesh. An accuracy of 1% is obtained from only 2,793 mesh nodes using the analytical field solution. This accuracy improves to 0.068% for 10,961 mesh nodes. The errors produced by the first-order finite element solution in comparison, are typically one to two orders of magnitude larger. It can be concluded that for the calculation of both third-order and fifth-order spherical aberration coefficients by direct ray tracing, interpolation errors can be safely kept below 0.1% on modest mesh sizes.

Number of nodes	% error on C_{s5} from analytical nodal potentials	% error on C_{s5} from first-order FEM solution
725	13.54	140
2,793	1.09	31.3
10,961	0.068	10.6
43,425	0.022	2.77

Table 8.3 Percentage accuracy of C_{s5}

2.4 Truncation errors

Tables 8.2b and 8.3 show that the accuracy on C_{s3} and C_{s5} using first-order finite element potential solutions is dominated by mesh truncation errors. It is of course possible to reduce the truncation error by increasing the mesh resolution. But since interpolation errors will also correspondingly decrease with mesh spacing, the overall error will still be dominated by truncation noise. However, other options exist. Table 8.4 presents results for second-order elements and extrapolated first-order element potential solutions on the region block mesh. By comparing these results to the ones presented in Tables 8.1-8.3, it is clear that the errors on C_{s3} and C_{s5} on modest mesh sizes are close to the limits set either by high-order interpolation or trajectory integration.

Number of nodes	% error on C_{s3} from extrapolation on first-order FEM solutions	% error on C_{s3} from a second-order FEM solution	% error on C_{s5} from extrapolation on first-order FEM solutions	% error on C_{s5} from a second-order FEM solution
725	6.68	62.0	20.5	181
2,793	0.32	1.44	2.11	19.57
10,961	0.024	0.078	0.15	1.13

Table 8.4 Percentage errors on C_{s3} and C_{s5} for extrapolation on first-order element and second-order element solutions

In summary, second-order elements are more accurate than first-order elements, but less accurate than first-order elements when extrapolation is used. The main point emerging from these results is that either by using extrapolation or high-order elements, truncation noise can be reduced to levels that are comparable to errors generated by field interpolation or trajectory integration for modest mesh sizes. For a mesh having less than 15,000 nodes, the errors on C_{s3} and C_{s5} can be reduced to around 0.01% and 0.1% respectively.

REFERENCES

[1]   Private correspondence from E. Munro, Mebs, 14 Cornwall Gardens, London SW7 4AN, UK

[2]   B. Lencová, "Compuation of electrostatic lenses and multipoles by the first-order finite element method", Nuclear Instruments and Methods in Physics Research A, 363, 1995, pp190-197

[3]   X. Zhu, H. Liu, J. Rouse and E. Munro, "Field function evaluation techniques for electron lenses and deflectors", Proc. SPIE Charged Particle Optics III, 1997, p68-80

[4]   B. Lencová, "Computation of electrostatic lenses and multipoles by the first-order finite element method", Nuclear Instruments and Methods in Physics Research A, 363 (1995), p190-195

[5]   X. Zhu and E. Munro, "Second-order finite element method and its practical application in charged particle optics", Journal of Microscopy, vol. 170, Pt. 2, August 1995, pp170-180

[6]   A. Khursheed, KEOS: The Khursheed Electron Optics Optics Software, Electrical Engineering Department, National University of Singapore, Singapore 119260, 1998

[7]   A. Khursheed and A. R. Dinnis, "High accuracy electron traectory plotting through finite-element fields", J. Vac. Sci. Technol. B7 (6), Nov/Dec 1989, p1882-5

[8]   L. Lapidus and J. H. Seinfeld, Numerical Solution of Ordinary Differential Equations, Academic, London, 1971, p68-9

[9]   A. Renau D. W. O Heddle, "Geometric aberrations in electrostatic lenses: 1. A simple and accurate computer model", J. Phys. E: Sci. Instrum. , vol. 19, 1986, p284-7

Chapter 9
MAGNETIC LENSES

There are several different ways in which the finite element method can be formulated for magnetic lens calculations. Munro's original formulation used the vector potential A(r,z) directly and employed first-order triangle elements [1]. He made a single point integration approximation to evaluate stiffness matrix terms. Lencová suggested improvements to Munro's orginal work [2]. She found that quadratic integration using triangle mid-side points improved the final accuracy, and that it was advantageous to reformulate the problem in terms of the potentials, $P(r,z)=2A(r,z)/r$ and $\psi(r,z)=2\pi rA(r,z)$. Lencová found that by employing graded mesh layouts, as opposed to Munro's region block meshes, the continuity of the axial field distribution was significantly improved. By using a Taylor's series expansion approach, she demonstrated that the finite element solution based upon the $P(r,z)$ and $\psi(r,z)$ potentials are less sensitive to discontinuities in mesh spacing than the original vector potential A(r,z). Lencová also carried out work on coupling the finite element method to the boundary integral method for open problems [3], as well as proposing that Ampere's circuital theorem be used to monitor the accuracy of axial field distributions [4]. Munro later used second-order elements and claimed that both in terms of field accuracy and continuity, they are superior to first-order order elements.

In this chapter, some simple test examples will be presented to compare the performance of first-order, second-order and third-order elements. Apart from the low-order integration schemes already tried by Munro and Lencová for first-order triangle elements, higher-order integration methods on quadrilateral first-order elements are investigated. The study reveals that increasing the order of the numerical integration scheme improves the accuracy and continuity of the axial field distribution. Quadrilateral first-order elements can give axial field distributions that are comparable in continuity to those produced by second-order elements, even on a mesh which has discontinuities in mesh spacing. In terms of accuracy, the results show that for a given amount of computation time, high-order elements perform better than first-order elements. This advantage is however, dependent on the way the problem is formulated and whether extrapolation is used. For the vector potential A(r,z), third-order elements are recommended. When using the potential P(r,z), it is better to use second-order elements. Where second-order elements are used with A(r,z), the accuracy can be significantly improved by using extrapolation.

1   ACCURACY ISSUES

1.1   The First-order approximation

The vector potential A(r,z) is an odd function, with A=0 on the optic axis,

$$A(r,z) = c_1 r + c_3 r^3 + c_5 r^5 + \ldots\ldots \tag{9.1}$$

Here, the coefficients, c_1, c_3....are functions of z, and the relationship between them is found from substituting the above expansion into the governing magneto-static differential equation in terms of A(r,z) (see Chapter 2, section 2). It is straightforward to show that the axial flux density $B(z)=2c_1(z)$ (equation 2.12).

The calculation for the axial field B(z) typically utilizes two off-axis values for A(r,z) along mesh lines in the radial direction. These mesh lines are assumed to be perpendicular to the optic axis. Let the mesh spacing in the radial direction be given by h, and let A_1 be the value of A(h,z) and A_2 be the value of A(2h,z). The value of B(z) can then be found by fitting these values to the first two terms in the expansion for A(r,z) and is given by $(8A_1-A_2)/3h$.

The governing magnetostatic equation (2.9) can be expressed as

$$A = r^2\left(\frac{\partial^2 A}{\partial z^2} + \frac{\partial^2 A}{\partial r^2}\right) + r\frac{\partial A}{\partial r} \tag{9.2}$$

Let the mesh spacing in the r and z directions be h and the centre node of a 5pt star be located at a radius r. It is straightforward to show that if ΔA is the truncation error on A(r,z), then the 5pt finite difference approximation in (r,z) is given by

$$A = \frac{r^2}{h^2}\left[A(r,z+h) + A(y,z-h) + \left(1-\frac{h}{2r}\right)A(r-h,z) + \left(1+\frac{h}{2r}\right)A(r+h,z)\right] + \Delta A$$

$$\tag{9.3}$$

Accounting for higher order terms in the Taylor's series leads to

$$\Delta A \approx \frac{h^2 B'' r}{16} - \frac{3h^2 B^{(IV)} r^3}{32} \tag{9.4}$$

Using the values of A(r,z) at r=h and r=2h, the error on the axial flux density is approximately given by

$$\Delta B \approx \frac{h^2 B''}{8} \tag{9.5}$$

The accuracy of the axial flux density from first-order finite element solutions in the vector potential A(r,z) should in principle be approximately given by this expression. The above finite difference error is similar to the second term in the expansion of the potential $P(r,z)$ evaluated for $r=h$. The Taylor's series in $P(r,z)$ is derived directly from the one used for $A(r,z)$ and is given by

$$P(r,z) = B - \frac{B''}{8}r^2 + \frac{B^{(IV)}}{192}r^4 + \ldots\ldots \tag{9.6}$$

Obviously for first-order elements formulated in terms of $P(r,z)$, the error ΔP for a distance h in the r direction will be of the order $h^2 B''/8$, identical to the error predicted by finite difference. Although first-order finite element solutions should in principle be similar to those obtained by 5pt finite difference, there are reasons why the finite element solution may be more inaccurate. Consider the finite element nodal equation for the magnetostatic problem formulated in the vector potential $A(r,z)$ (equation (7.6)). In the expression for K_{ij}, there are singular terms such as $N_i N_j / r$

$$K_{ij} = 2\pi \iint_{\Omega_e} \frac{1}{\mu}\left[\left(\frac{\partial N_i}{\partial z}\right)\left(\frac{\partial N_j}{\partial z}\right) + \left(\frac{N_i}{r} + \frac{\partial N_i}{\partial r}\right)\left(\frac{N_j}{r} + \frac{\partial N_j}{\partial r}\right)\right] r\,dr\,dz \tag{9.7}$$

The order of numerical integration is important. Some common numerical integration schemes for first-order elements are shown in Figure 9.1. The exact coordinates for the integration points are given in Appendix 1. It turns out that the high-order integration scheme, that is the 6 pt by 6pt Gaussian Quadrature method on the quadrilateral element, provides results which are comparable to 5pt finite difference and yields more continuous axial field distributions.

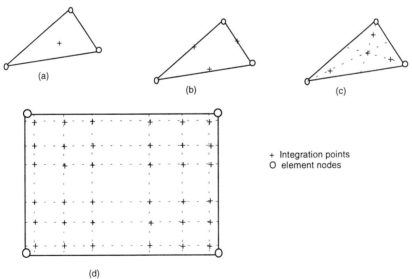

Figure 9.1 Some integration schemes on first-order elements
(a) Triangle 1pt (b) Triangle 3pt integration (c) Triangle 4pt
(d) Quadrilateral 6pt by 6pt

1.2   The solenoid example

Consider the solenoid coil shown in Figure 9.2. Let the coil lie between a_1 and a_2 in the radial direction, and between z_1 and z_2 in the longitudinal direction z.

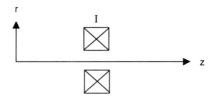

Figure 9.2 The Solenoid layout

The analytical solution for the solenoid field can be found from the expression for current loops [5]. The vector potential A(r,z) for a current loop is

$$A(r,z) = \frac{\mu_0 I}{\pi k}\left(\frac{a}{r}\right)^{1/2}\left[\left(1-\frac{k^2}{2}\right)K(k^2)-E(k^2)\right] \qquad k^2 = \frac{4ar}{z^2+(a+r)^2} \qquad (9.8)$$

where a is the loop radius, I is the loop current (in the azimuthal direction), K and E are elliptic integrals of the first and second kind respectively. The vector potential for the solenoid is found by integrating over the coil region. Near the axis, in the paraxial region, the above formula is not suitable and the elliptic integrals must be replaced by series expansions (see Appendix 4). Note that the components of the flux density can readily be found in either case.

The simulation model for the solenoid is illustrated in Figure 9.3. An axis of even symmetry lies at z=0, and the boundary conditions for the vector potential at the outer boundaries $A_B(r,z)$, are derived from the above formula. The inner and outer radii of the coil are given by 10 and 20 mm respectively, and its length in the z direction extends from –5mm to 5mm. A current of 100 AT is assumed. A very simple mesh layout was specified where the mesh spacing in the radial direction is taken to be 1 mm. In the z-direction, a mesh spacing of 0.833 mm is used. The minimum number of mesh lines in both r and z directions is 61 which spans an area of 50 by 50 mm. The mesh resolution was successively doubled and quadrupled to obtain finer mesh lines. The flux lines for a typical finite solution of this problem is shown in Figure 9.4.

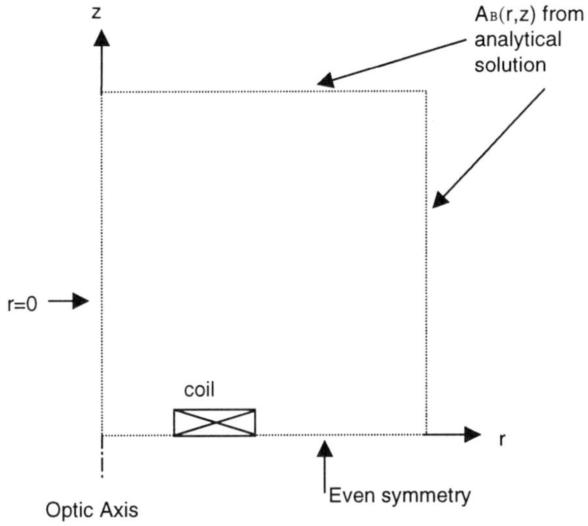

z

A$_B$(r,z) from
analytical
solution

r=0 →

coil

r

Even symmetry

Optic Axis

Figure 9.3 Simulation model for solenoid test example

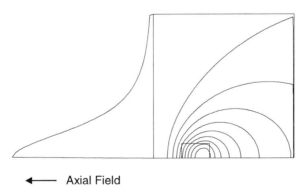

← Axial Field

Figure 9.4 Flux lines for the solenoid test example

1.3   Axial field errors for first and second-order elements

The normalised axial field distribution and its first two derivatives according to the analytical formula are shown in Figure 9.5. A peak value of around 50 T/m^2 was obtained for the second derivative at z=0. These graphs are helpful in understanding the errors associated with the finite element solution.

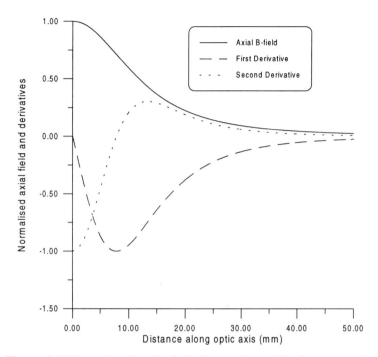

Figure 9.5 The Solenoid axial field distribution and its first two derivatives

The first set of results generated for the solenoid problem are formulated in terms of the vector potential A(r,z). Figure 9.6a shows the axial field error as a function of z obtained for first-order triangles using 3pt integration (quadratic), while Figure 9.6b shows the results for 9 node second-order finite elements. The first point to make about the first-order element results is that the error is proportional to the second derivative B″(z). The second observation is that the error from first-order triangle elements is approximately given by $h^2B''/3$ (peak value of 16 micro-Tesla), a factor of around 2.5 larger than that predicted for 5 pt finite difference ($h^2B''/8$). Munro's orginal method (triangle element 1pt integration) gives a peak error of around 18 micro-Tesla. The third observation is that for both first and second-order elements, the error is proportional to h^2, this means that Richardson's extrapolation will be an effective method for improving the accuracy. The fourth observation is that the error for second-order elements is around a factor of 13 times smaller than that for first-order elements on the same mesh.

To give comparable errors, the mesh spacing of the first-order element program needs to be around ¼ of that used for the second-order program. This means to provide the same accuracy, first-order triangle elements require approximately 16 times more mesh nodes than second-order elements. But since the program execution times for second-order elements were found to be around 15 times longer than first-order elements on the same mesh, they are comparable in performance. The results are summarized in Table 9.1.

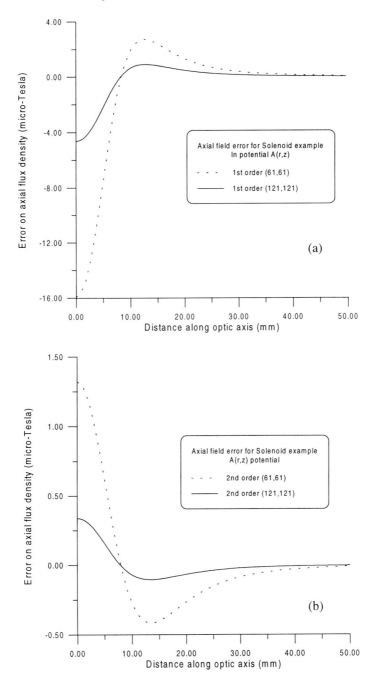

Figure 9.6 Solenoid axial field error in A(r,z)
  (a) First-order triangle elements using 3pt integration
  (b) 9 node second-order elements

In A(r,z) potential	First-order triangle Elements	9-node Second-order Elements
Peak ΔB on (61,61) mesh (micro-Tesla)	-15.9	1.31
Peak ΔB on (121,121) mesh (micro-Tesla)	-4.64	0.337
Peak ΔB Extrapolated value (micro-Tesla)	-0.88	0.0097
Run time for (61,61) mesh (CPU seconds)	0.65	9.78
Run time for (121,121) mesh (CPU seconds)	2.47	37.7

Table 9.1 Axial field accuracy for first-order triangle elements with 3pt integration and 9 node second-order elements solved in terms of A(r,z)

From the above results, it is clear that extrapolation is more effective for second-order elements. In both cases, extrapolation significantly reduces the error. An extrapolated field value for second-order elements is around two orders of magnitude more accurate than the comparable first-order element value.

Figures 9.7a and 9.7b show graphs of the axial field distribution error for the potential P(r,z). The results are summarized in Table 9.2

In P(r,z) potential	First-order Elements	Second-order Elements
Peak ΔB on (61,61) mesh (micro-Tesla_	-3.45	-0.6
Peak ΔB on (121,121) mesh (micro-Tesla)	-1.04	-0.043
Peak ΔB Extrapolated value (micro-Tesla)	0.65	-0.0055
Run time for (61,61) mesh (CPU seconds)	0.7	10.16
Run time for (121,121) mesh (CPU seconds)	3.1	38.24

Table 9.2 Axial field accuracy for first-order triangle elements with 3pt integration and 9 node second-order elements solved in terms of P(r,z)

The results for finite element calculations in P(r,z) exhibit significant improvement over using A(r,z). For first-order triangle elements using 3pt inetgration, they are around a factor of four times more accurate than in A(r,z). These results are evident from comparing Figures 9.6a and 9.7a, and are also depicted in Figure 9.8.

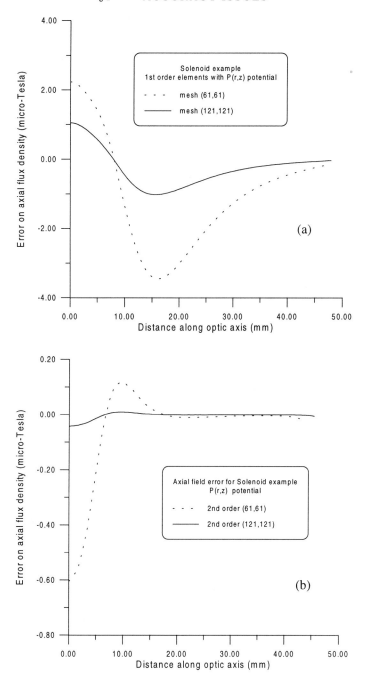

Figure 9.7 Axial field error in P(r,z)
(a) First-order triangle elements using quadratic integration
(b) 9 node second-order quadrilateral elements

Although 4pt (cubic) integration for triangle elements in $P(r,z)$ was found to be more accurate than 3pt (quadratic) integration, the factor of improvement is not sustained along the whole optic axis. Extrapolation is not as effective when using $P(r,z)$. The final result after using extrapolation is the same whether 3pt or 4pt integration is used. More importantly, after extrapolation, the accuracy for the solution in $A(r,z)$ is approximately equal to the accuracy of the solution in $P(r,z)$. This result is shown in Figure 9.9. So for first-order elements, although the accuracy is improved by reformulating the potential in $P(r,z)$, after extrapolation, no advantage is gained.

The results for quadrilateral first-order elements formulated in $A(r,z)$ using 6pt by 6pt integration in either direction gives virtually the same performance as 5pt finite difference, as shown in Figure 9.10. The extrapolated value in this case gives very similar results to the other first-order element methods. In fact, after extrapolation, all the first-order elements investigated gave around the same level of accuracy as obtained by 5pt finite difference, including Munro's original formulation.

Table 9.2 indicates that for second-order elements in $P(r,z)$, the factor of improvement depends on the mesh resolution. This is because the improvement in accuracy depends on h^4, and not on h^2: on the (121,121) mesh, the peak error for the $P(r,z)$ potential is approximately one order of magnitude better than for the one derived from the $A(r,z)$ potential. However, less than a factor of two improvement is obtained by using $P(r,z)$ after extrapolation as opposed to using $A(r,z)$. This is also illustrated in Figures 9.11 which shows the axial field error distributions for $A(r,z)$ and $P(r,z)$ after extrapolation. Similar to the situation with first-order elements, although the errors associated with the potential $A(r,z)$ are greater than that for $P(r,z)$, their variation with h is more predictable, so after extrapolation, the results from $P(r,z)$ and $A(r,z)$ give the same level of accuracy.

After extrapolation, second-order-element axial field values are around two orders of magnitude more accurate than those obtained from first-order elements. Since second-order elements typically take between 11 to 15 times longer to execute than first-order elements, it can be concluded that there is an overall advantage in using second-order elements as opposed to using first-order elements. If extrapolation is not used, the accuracy of second-order elements are comparable to first-order elements if potential $A(r,z)$ is used. However, in this case there is a definite advantage in formulating the problem in potential $P(r,z)$. In $P(r,z)$, first-order elements perform better because they avoid the $1/r$ singularity in the stiffness element integral. For second-order elements, the $P(r,z)$ potential is better because the error is dominated by radial variations of the potential, and since $P(r,z)$ varies with even powers of r, second-order elements are better suited to model the radial variations of $P(r,z)$. The potentials rA and $A/r^{1/2}$ were tried with second-order elements, but they yielded inaccurate results. Further work is required to investigate their accuracy performance.

To summarise: second-order elements are more accurate than first-order elements; if extrapolation is not used, $P(r,z)$ gives more accurate results than using $A(r,z)$; if extrapolation is used, the level of accuracy from $A(r,z)$ and $P(r,z)$ are similar.

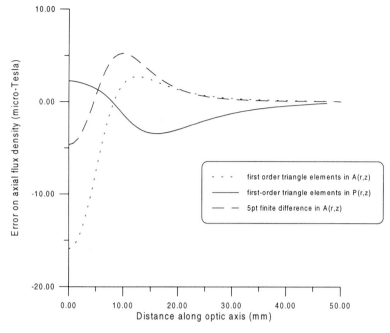

Figure 9.8 Comparison of first-order triangle element solution in A(r,z) and P(r,z) on the (61,61) mesh

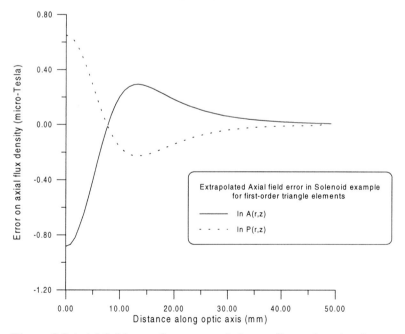

Figure 9.9 Axial field error from extrapolation on first-order triangles

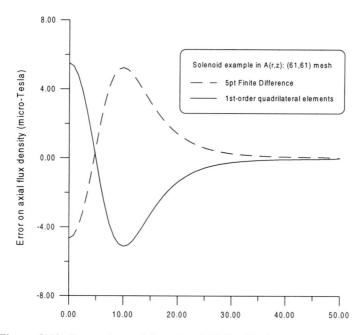

Figure 9.10 Comparison of the axial field distribution error from quadrilateral elements with 5 pt finite difference

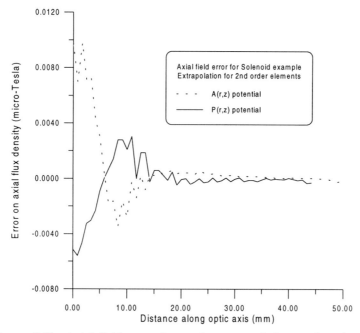

Figure 9.11 Axial field error from using extrapolation on 9 node second-order element solutions

1.4   Cubic elements

To examine the accuracy of cubic elements, serendipity elements were used. From considerations of the Taylor's series expansions of $P(r,z)$ and $A(r,z)$ (see equations (9.1) and (9.6)), cubic elements are expected to be better suited to the vector potential $A(r,z)$, since the potential in this case is proportional to odd powers of r. For $P(r,z)$, which depends on even powers of r, second-order elements are expected to perform better than third-order elements. These expectations are confirmed by the results depicted in Figures 9.12 and 9.13 which compare the accuracy performance of third-order elements to second-order elements for the potentials $A(r,z)$ and $P(r,z)$ respectively. The accuracy comparison is made as a function of mesh node number and relative computation time.

From the results for accuracy as a function of mesh node number, shown in Figures 9.12a and 9.13a, it appears that third-order elements are more accurate than second-order elements for both $A(r,z)$ and $P(r,z)$. This is the usual kind of comparison carried out in finite element literature and it is incorrect. This is because the number of non-zeros entries in the final matrix rises as the element order increases, reflecting the fact that there is greater coupling between the mesh nodes for high-order elements: a greater number of surrounding nodes are involved in the nodal equation for higher-order elements. A better comparison is to examine the accuracy as a function of total computation time. Since serendipity elements are used, this is easily achieved. Figures 9.12b and 9.13b show that the accuracy of third-order elements is in fact better than second-order elements for the potential $A(r,z)$, as expected, but do not perform as well for the potential $P(r,z)$. In fact for a given amount of computation time, when formulated in the potential $P(r,z)$, second-order elements are able to provide better accuracy than third-order elements. Figure 9.14 confirms that second-order serendipity elements are much more accurate when formulated in $P(r,z)$. Since extrapolation was not found to be effective for the $P(r,z)$ potential, it is recommended that if third-order elements are used, they be used in the vector potential $A(r,z)$.

To summarize: third-order elements are more accurate than second-order elements for the potential $A(r,z)$, while for the potential $P(r,z)$, second-order elements provide better accuracy.

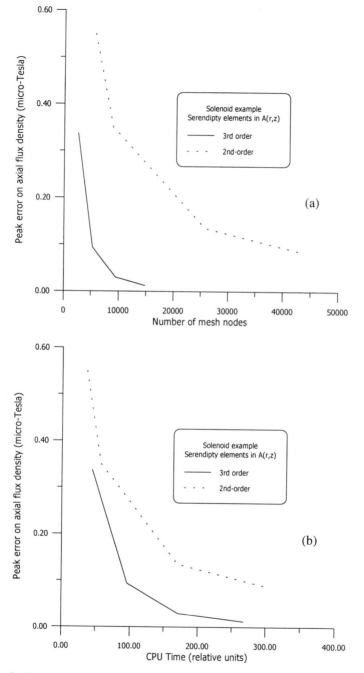

Figure 9.12 Comparison of the axial field accuracy from third and second-order
serendipity elements
(a) As a function of the number of mesh nodes
(b) As a function of computation time

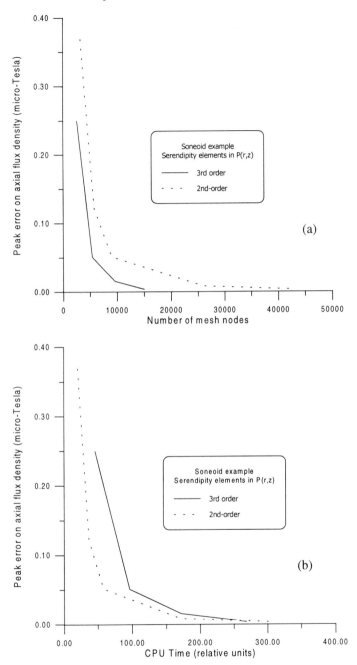

Figure 9.13 Comparison of the axial field accuracy from third and second-order
serendipity elements
(a) As a function of the number of mesh nodes
(b) As a function of computation time

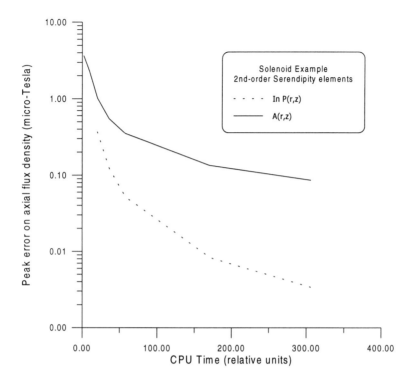

Figure 9.14 Comparison of the axial field accuracy of second-order serendipity elements for potentials A(r,z) and P(r,z)

2.   MAGNETIC AXIAL FIELD CONTINUITY TESTS

2.1   First-order elements on a trial region block mesh

To investigate axial magnetic field continuity, finite element field calculations were made for the Cambridge Instruments Scanning Electron Microscope S100 objective lens. The diagrams of a region block mesh and a typical field distribution generated for this lens are shown in Figures 9.15a-c. Since a variety of different mesh types were used, this block region mesh will be referred to here as the "trial mesh". In the following study, the continuity of the third derivative of the axial field distribution will be examined. Since the axial flux density is everywhere in air, all its higher derivatives should be continuous. Higher derivatives of the axial field distribution were found by using cubic spline interpolation.

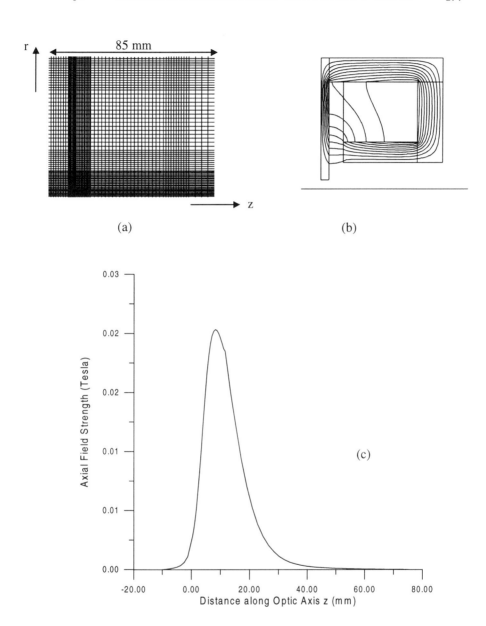

Figure 9.15 S100 Objective Lens
  (a) Mesh
  (b) Flux lines
  (c) Axial field distribution

Figure 9.16a shows the results for a first-order triangle element program using 3pt integration formulated in the vector potential A(r,z). The third derivative is obviously highly discontinuous. On the other hand, much of the discontinuity can be eliminated by reformulating the problem in terms of the potential P(r,z) and 4pt (cubic) integration. This result is shown in Figure 9.16b and is similar to the axial field third derivative from a second-order finite element program, shown in Figure 9.16c. Although slightly smoother than the first-order element solution, the second-order element program took over 12 times longer to run.

The better continuity in P(r,z) for first-order triangle elements is critically dependent on the order of numerical integration. Using 3pt integration did not provide significant improvement. Improvement was obtained only when 4pt integration or higher was used in combination with the potential P(r,z). Also, as an additional check, 4pt and 7pt integration was used with the original potential A(r,z), but this did not significantly change the final result.

Another important result relates to using first-order quadrilateral elements with high-order integration (6pt by 6pt Gaussian Quadrature). Here a smooth derivative, similar to the one produced by 9 node second-order elements is produced for the problem formulated in A(r,z). The results are shown in Figure 9.16d. They show that it is not necessary to re-formulate the problem in a different potential other than A(r,z)) in order to improve the continuity of the axial field distribution of first-order element solutions. This result shows that the poor continuity performance of Munro's original formulation is not an intrinsic feature of first-order elements but is related to its low-order integration method.

The foregoing results show that for the test example considered, first-order elements can provide comparable axial field continuity to second-order elements. To consolidate this conclusion, a second test example was selected, one which has previously been used to demonstrate that first-order elements gives discontinuous results. Figures 9.17a-c show the mesh layout, flux lines and axial field distribution of an objective lens presented by Zhu et al [6]. They presented axial field derivatives up to third order which demonstrate that second-order elements provide much better continuity than first-order elements. Note that the lens is operating in the saturation region. In the present work, it was found that while the original Munro program gives a highly discontinuous axial field third derivative, first-order quadrilateral elements using 6pt by 6pt Gaussian Quadrature integration yields smooth results, and provides an axial field third derivative which is similar to the one produced by second-order elements. These results are shown in Figures 9.18a and 9.18b.

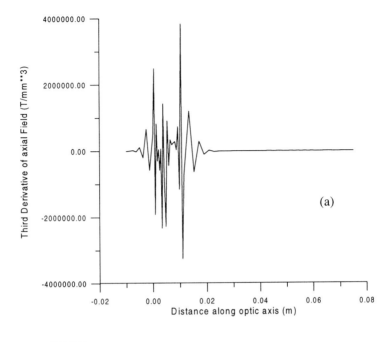

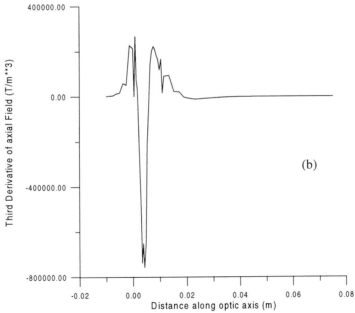

(Figure 9.16)

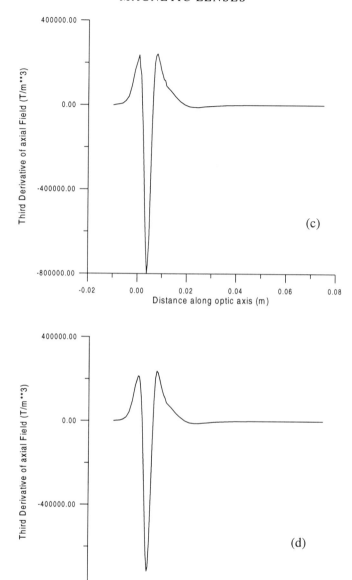

Figure 9.16 Third-order derivative of the axial field distribution solution on the trial mesh of the S100 Lens
(a)   First-order triangle elements in A(r,z) with 3pt integration
(b)   First-order triangle elements in P(r,z) with 4pt integration
(c)   9 node second-order elements in A(r,z)
(d)   First-order quadrilateral elements with 6pt by 6pt integration

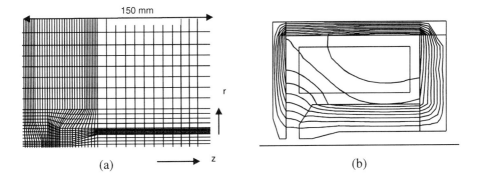

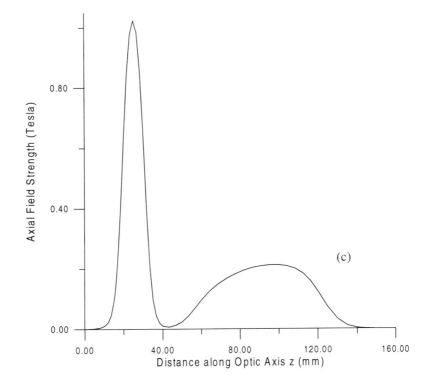

Figure 9.17 Test Saturated Objective Lens
(a) Mesh lines   (b) Flux lines and axial field distribution
(c) Axial Field Distribution

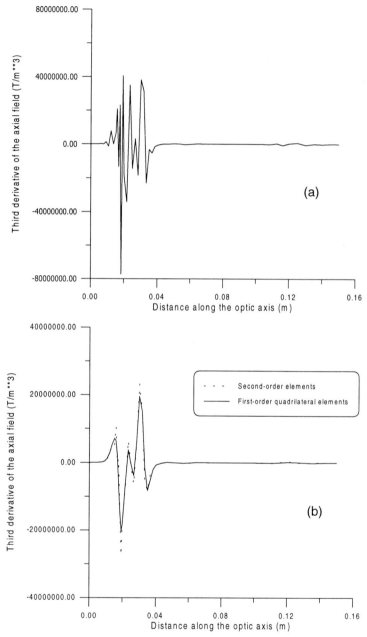

Figure 9.18 Third derivative of the axial field distribution for the test magnetic
saturated lens
(a) First-order triangle elements in A(r,z) using 3pt integration
(b) Second-order and first-order quadrilateral elements with 6 pt by 6pt integration

2.2   Mesh Refinement

One way of improving the axial field continuity of Munro's original program, as suggested by Lencová, is by grading the mesh [7]. Instead of using the trial region block mesh shown in Figure 9.15a for the S100 lens, the mesh was graded, as shown in Figure 9.19a. In the gap region, constant mesh spacing is used along the optic axis, while for regions outside it, the mesh was gradually expanded by a factor of 1.02 from mesh cell to mesh cell. The result on the axial field distribution, depicted in Figure 9.19b, shows that even for Munro's original implementation of first-order finite elements in the vector potential A(r,z), the discontinuity of the axial field distribution can largely be eliminated.

Another form of mesh refinement is to construct a quasi-conformal mesh. This method has been proposed by Khursheed and is described in Chapter 11, section 1, and has obvious advantages over grading the mesh. Here, since the aspect ratio of every mesh cell is always close to unity, truncation and interpolation errors are minimised. This type of mesh is also convenient for direct ray tracing of secondary electrons, since higher interpolation on such a structured orthogonal meshes can be fast and accurate.

The quasi-conformal mesh is formed by overlaying the equi-potential lines of two conjugate potential distributions defined in Cartesian coordinates. A quasi-conformal refinement mesh for the S100 lens is depicted in Figure 9.20a. The magnetic field solution is now re-solved on this refined mesh. The third derivative of the axial field distribution, found by using first-order triangle elements with single point integration is shown in Figure 9.20b. It is clear that the discontinuities of the trial solution have been eliminated, and that quasi-conformal mesh refinement significantly improves the continuity of the lens axial field distribution.

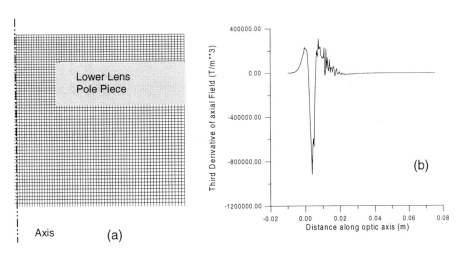

Figure 9.19 Graded Mesh Refinement for the S100 Lens
(a) Refinement Mesh   (b) Third derivative the axial field distribution calculated from first-order triangle elements in vector potential A(r,z)

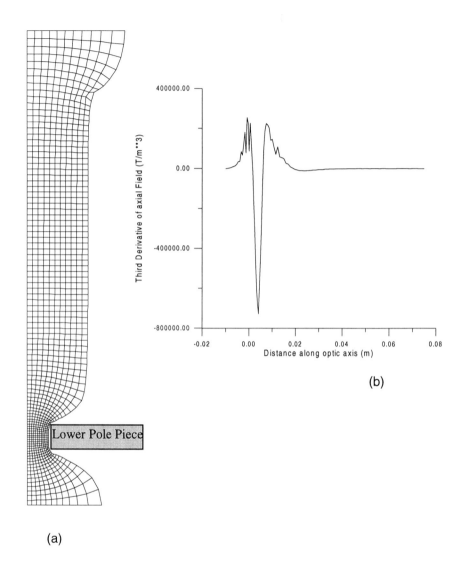

(a)

Figure 9.20 Quasi-conformal Mesh Refinement for the S100 Lens
(a) Refinement Mesh
(b) Third derivative the axial field distribution calculated from first-order triangle
     elements in vector potential $A(r,z)$

2.3   Direct Ray tracing from axial field distributions

The foregoing results can be readily quantified by using direct ray tracing to determine on-axis third and fifth order spherical aberration coefficients, C_{s3} and C_{s5} respectively. Here the axial flux density and its derivatives up to third order are used to plot electron trajectories. The Taylor expansions of the field components $B_r(r,z)$ and $B_z(r,z)$ in terms of the axial flux density $B_0(z)$, are used directly in the time-dependent equations of motion. The following equations are solved

$$m\ddot{x} = -e\left(v_y B_z - v_z B_y\right) \quad m\ddot{y} = -e\left(v_z B_x - v_x B_z\right) \quad m\ddot{z} = -e\left(v_x B_y - v_y B_x\right) \quad (9.9)$$

where terms up to the third derivative of $B_0(z)$ are retained

$$B_z(r,z) \approx B_0(z) - \frac{1}{4}\frac{\partial^2 B_0(z)}{\partial z^2}$$

$$B_r(r,z) \approx -\frac{1}{2}\frac{\partial B_0(z)}{\partial z}r + \frac{1}{16}\frac{\partial^3 B_0(z)}{\partial z^3}r^3 \qquad (9.10)$$

$$B_x(r,z) = B_r(r,z)\cos\theta \quad B_y(r,z) = B_r(r,z)\sin\theta \quad \tan\theta = \frac{y}{x}$$

Note that this method of direct ray tracing is different to the one used in Chapter 8, section 2, where direct ray tracing is calculated from off-axis potentials. Here, direct ray tracing uses axial field expansions. A simple 4th-order Runga-Kutta technique to plot the trajectories is used, and a two parameter least squares fit on the change of the focal point Δz with C_{s3} and C_{s5} from a set of incoming parallel trajectories is made. This procedure is similar to the one used for calculating spherical aberration coefficient information by direct ray tracing for electric lenses in chapter 8. The test example to be examined here is the S100 Objective Lens shown in Figure 9.15.

It is primarily the 5th order spherical aberration coefficient that will be examined here. This is because the axial field first derivative is important for estimating the 3rd order spherical aberration coefficient, which can be calculated sufficiently accurately by first-order finite elements on a region block trial mesh for most problems. An indication that this is the case can be seen by comparing the estimate of C_{s3} by direct ray tracing with the value calculated from perturbation of the paraxial trajectory. For the S100 objective lens region block mesh shown in Figure 9.15a, the result by perturbation is 2.05 cm, while from direct ray tracing it is typically 1.98 cm, giving an agreement of around 3.5%. For this direct ray tracing value, only 11 trajectories were plot over an aperture radius of 0.5 mm. Better agreement was obtained for more trajectories and smaller radii.

For the calculation of the spherical fifth-order aberration coefficient, it is the input radii of the electron trajectories which is the critical parameter. The further off-axis they are, the more important is the effect of axial field discontinuities, as seen from equation (9.10). For this study, 200 trajectories are plot within an input radius of 4.4 mm. This aperture radius lies just within the lens bore, which has a radius of 4.5 mm. At this radius, it is primarily fifth-order aberrations which dominate the size of the beam at the specimen. The calculated values for C_{s5} are summarised in Table 9.3.

Direct Ray Tracing	C_{s5} (cm)
Triangle first-order elements on the trial mesh with 3pt integration (65,66) (129,131) Extrapolated value	 -40.9 -24.6 -30.3
Quadrilateral elements with 6 pt by 6pt integration on the trial mesh (65,66)	 -5.5
Second-order elements on the trial Mesh (65,66) (129,131) Extrapolated value	 -5.39913 -5.399089 -5.3861
Quasi-Conformal Mesh Refinement (12,100) (23,199) Extrapolated value	 -5.46292 -5.38688 -5.37609
Graded Mesh (82,121) (163,241) Extrapolated value	 -5.368 -5.3709 -5.3728

Table 9.3 A comparison of the fifth-order on-axis spherical aberration coefficient derived by direct ray tracing on the S100 objective lens for various finite element schemes

Although the true value of C_{s5} is not known, accuracy here is taken to be a measure of how well different finite element schemes converge. After extrapolation, the second-order element C_{s5} value on the trial mesh gives good agreement with the values found from first-order element solutions on the graded and quasi-conformal meshes. The graded and quasi-conformal mesh values of C_{s5} agree to within 0.063%, while their agreement with the second-order element value is less than 0.25%. These results show that accurate values of C_{s5} can be obtained either through refining the order of the elements or by mesh refinement. On the hand, first-order triangle element 1pt integration solutions (Munro's original formulation) on the trial mesh are clearly very inaccurate. For first-order quadrilateral elements which employ 6pt by 6pt Gaussian Quadrature integration, a C_{s5} value of –5.5 cm is obtained, giving an agreement of around 2.3% with the second-order element value. This shows that even on the trial mesh, if properly formulated, first-order elements can provide accurate estimates of C_{s5}. These results confirm the continuity results of the previous section. They show that poor axial field continuity is not an intrinsic feature of first-order elements, but due to relatively low-order integration schemes.

3.   MAGNETIC FIELD COMPUTATION IN THREE DIMENSIONS

This section compares the accuracy of the reduced scalar potential and two-scalar potential methods in calculating three dimensional magnetic field distributions. A finite difference implementation of the reduced scalar potential method has been used by Rouse and Munro [8], while the more general two scalar potential formulation is available from the company Vector Fields [9]. The main disadvantage of the reduced scalar potential approach, namely that it has field cancellation errors, needs to be tested on problems relevant to charged particle optics.

An obvious situation where field cancellation is expected to generate large errors is for the calculation of stray fields. Consider for instance the penetration of magnetic fields through the walls of an iron chamber, such as the specimen chamber of an electron microscope. The magnetization field and source field will obviously be of the same order and opposite sign, causing the fields inside the chamber to be small. This is a difficult case even for two dimensional field solving programs, since in general, truncation errors greatly exceed the strength of stray fields.

Another possible source of error for the reduced scalar potential approach arises for the situation where the specimen is placed just above a pole-piece which lies on axis, such as the single pole-piece lens advocated by Mulvey [10]. The key issue here is to establish where the cancellation errors occur. If they are located inside iron regions, then they do not pose a problem since the primary beam only propagates in free-space. The S100 objective lens (shown in Figure 9.15b) is used once again as a test example, only in this case, the gap in the lens lower pole-piece is removed, as shown in Figure 9.21. The axial region of interest ranges from 0 to 5 mm above the lower pole-piece.

Since this problem has axisymmetry, the three dimensional field distribution can be compared to its two-dimensional counterpart. Figure 9.22 compares the axial field distribution obtained from a reduced scalar potential solution, total scalar potential solution, and axisymmetric vector potential solution. The two three-dimensional finite element programs employ first-order tetrahedral elements and were generated by TOSCA, a Vector Fields program, while the two-dimensional vector potential program uses second-order elements and was written by the author.

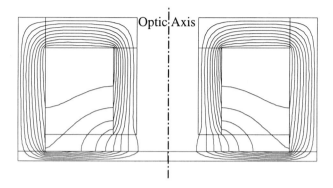

Figure 9.21 Modified S100 Objective Lens

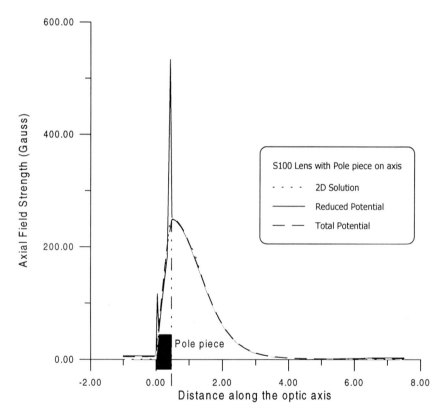

Figure 9.22 Modified S100 objective lens axial field distributions from two and three dimensional finite element solutions.

The three-dimensional mesh consists of region blocks in the longitudinal direction and a polar mesh in the transverse plane. The number of nodes in the longitudinal and radial directions were chosen to be similar to the 2D mesh model, while 16 mesh lines were evenly spaced over 2π radians in the azimuthal direction. The results show that fictitious spikes in the reduced scalar potential solution occur at the air/iron interfaces, while the two-scalar potential solution gives results which are very close to the axisymmetric solution. If a thin specimen, such as a 300 micron specimen is placed on the lower pole-piece, then the reduced scalar potential solution would give highly inaccurate results for the axial field at the surface of the specimen. The size of the first mesh above the pole-piece in the z direction is 561 microns. Even when the mesh resolution is doubled, significant spikes are obtained at the iron/air interfaces. For these types of situations, the use of the two-scalar potential method is recommended.

It should be noted that with the advent of edge finite elements, new possibilities exist for three dimensional magnetic field computations. Webb and Forghani propose a way of combining a single continuous scalar potential with edge elements which does not suffer from the loss of accuracy associated with the reduced scalar potential [11].

REFERENCES

[1]   E. Munro, Computer-Aided Design Methods in Electron Optics, PHD Thesis, University of Cambridge, England, October 1971

[2]   B. Lencová, "The use of the first-order FEM for the compuation of magnetic electron lenses", Proc SPIE, Charged Particle Optics II, Vol. 2858, 1996, p34-45

[3]   B. Lencová and M. Lenc, "The computation of open electron lenses by coupled finite element and boundary integral methods", Optik, Vol. 68, No. 1, 1984, p37-60

[4]   B. Lencová, "Unconventional lens computation", Journal of Microscopy, Vol. 179, Pt. 2, August 1995, p185-190

[5]   J. D. Jackson, Classical Electrodynamics, 2nd edition, John Wiley and Sons, NY, 1975, section 5.5, p177-8

[6]   X. Zhu, H. Liu, J. Rouse and E. Munro, "Field evaluation techniques for electron lenses and deflectors", Proc SPIE, Charged Particle Optics II, Vol. 2858, 1996, p68-80

[7]   B. Lencová, "The use of the first-order FEM for the compuation of magnetic electron lenses", Proc SPIE, Charged Particle Optics II, Vol. 2858, 1996, p34-45

[8]   J. Rouse and E. Munro, "Three-dimensional computer modelling of electrostatic and magnetic electron optical components", J. Vacuum Science and Technology, B7(6), 1989, p1891-1897

[9]   TOSCA, Vector Fields Ltd, 24 Bankside, Kidlington, Oxford OX5 1JE, England

[10] T. Mulvey, "Unconventional Lens Design", chapter 5 in "Magnetic Electron Lenses", editor P. W. Hawkes, Springer-Verlag, New York, 1982

[11] J. P. Webb and B. Forghani, "A single scalar potential method for 3D magnetostatics using edge elements", IEEE Transactions on Magnetics, Vol. 25, No. 5, September 1989, p4126-8

Chapter 10
DEFLECTION FIELDS

In chapter 2, deflections systems were briefly described in differential form. Here they are given a finite element formulation. The following section is based upon two papers by Munro and Chu on magnetic and electrostatic deflectors [1], and the work of Lencová [2]. The accuracy of the finite element method for deflectors will then be examined on some simple test examples. Much of the work presented in this chapter has been carried out by Zhao [3].

1.   FINITE ELEMENT FORMULATION

1.1   Energy functional

In chapter 2, it was shown that deflection systems in charged particle optics can be described by a scalar potential distribution $\Phi(r,z)$ satisfying the following divergence equation

$$\nabla.[\gamma(\nabla\Phi + k\mathbf{F})] = 0 \qquad (10.1)$$

where $\gamma=\varepsilon$, $k=0$ for electrostatic deflection fields, $\gamma=\mu$, $k=1$ for magnetic deflection fields and the vector function $\mathbf{F}$ is related to the driving current of deflectors such as saddle and toroidal coils.

The variational form of this equation is given by

$$U = \iiint_{\Omega} \frac{1}{2}\gamma \left|\nabla\Phi + k\mathbf{F}\right|^2 d\Omega \qquad (10.2)$$

The volume of integration extends over the whole domain Ω. Minimising this functional with respect to Φ, subject to the prescribed boundary conditions, corresponds to the solution of the differential equation (10.1).

The vector function $\mathbf{F}$ can be generally expressed as follows

$$\mathbf{F} = g(r,z)(a_1\mathbf{i}_r + a_2\mathbf{i}_z + a_3\mathbf{i}_\theta) \sum_{m=1,3,5,...}^{\infty} f_m \cos m\theta \qquad (10.3)$$

where the function $g(r,z)$ equals unity inside the windings and zero everywhere else, including electrode regions for electrostatic cases, and f_m is a driving function for the deflector which can be related to the driving current NI (see chapter 2, sections 4.3 and 4.4). For toroidal coils, $a_1=a_2=0$ and $a_3=1$, while for saddle coils $a_1=\cos\alpha$, $a_2=\sin\alpha$ and $a_3=0$, where α describes the taper angle. For electrostatic deflectors, the parameter k in equation (10.1) should be 0, and is equivalent to $g(r,z)=0$.

We now substitute the above expression for $\mathbf{F}$ into equation (10.2) and evaluate the θ integral from 0 to 2π. This yields the function U_m for the mth harmonic of the potential distribution, $\Phi_m(r,z)$

$$U_m = \frac{\pi}{2} \iint_\Omega \gamma \left[\left(\frac{\partial \Phi_m}{\partial r} + a_1 g f_m \right)^2 + \left(\frac{\partial \Phi_m}{\partial z} + a_2 g f_m \right)^2 + \left(-\frac{m\Phi_m}{r} + \frac{a_3 g f_m}{r} \right)^2 \right] r \, dr dz$$

$$(10.4)$$

where the region of integration Ω is the entire area of the (r,z) plane.

In order to avoid having to differentiate $\Phi_m(r,z)$ for the evaluation of the deflection function $d_m(z)$ on axis, Lencová uses the following modified expression for the harmonic potential function Φ_m

$$\Phi_m(r,z) = r^m \Psi_m(r,z) \qquad (10.5)$$

$\Psi_m(r,z)$ will be referred to as the modified harmonic potential. On the optic axis $d_m(z) = \Psi_m(0,z)$ (see equation 2.45). The advantages of using the function Ψ_m is that the radial dependence next to the axis is removed, and the explicit boundary condition $\Phi_m = 0$ on the optic axis is replaced by the Neumann boundary condition $\partial \Psi_m/\partial r = 0$ which is automatically satisfied in the energy functional.

Following Lencová's derivation, we substitute the above expression for Ψ_m in the energy functional for each harmonic component

$$U_m = \frac{\pi}{2}(L + R + A) \qquad (10.6)$$

where the expression for L is

$$L = \iint_\Omega \gamma r^{2m} \left[r \left(\frac{\partial \Psi_m}{\partial r} \right)^2 + r \left(\frac{\partial \Psi_m}{\partial z} \right)^2 + 2m\Psi_m \frac{\partial \Psi_m}{\partial r} + 2m^2 \frac{\Psi_m^2}{r} \right] dr dz \qquad (10.7)$$

and the expression of R is

$$R = 2gf_m \iint_\Omega \gamma r^m \left[a_1 \left(r\frac{\partial \Psi_m}{\partial r} + m\Psi_m \right) + a_2 r \frac{\partial \Psi_m}{\partial z} - a_3 m \frac{\Psi_m}{r} \right] drdz \qquad (10.8)$$

the last term a is given by

$$A = g^2 f_m^2 (a_1^2 + a_2^2 + \frac{a_3^2}{r^2}) r \qquad (10.9)$$

Note that A does not involve the harmonic potential Ψ_m, so that it will not appear in the finite element nodal equations.

For electrostatic deflectors, the modified harmonic potential on a rotational surface containing the electrode can be rewritten as

$$\Psi_B(z) = \frac{4}{\pi m^2 r(z)^m} \left\{ \left[\sum_{i=1}^{n-1} (V_{i+1} - V_i) \left(\frac{\cos m\theta_{a_{i+1}} - \cos m\theta_{b_i}}{\theta_{a_{i+1}} - \theta_{b_i}} \right) \right] + V_n \left(\frac{\cos m\theta_{b_n}}{\pi/2 - \theta_{b_n}} \right) \right\}$$

$$(10.10)$$

where $r(z)$ describes how the radius of the electrode boundary changes with z, n is the number of electrodes and the angular aperture of each electrode in the transverse plane (x-y plane) is defined by θ_{ai} to θ_{bi} (see Figure 2.12). The above equation is a modified form of equation (2.51). This expression applies generally to any rotational electrodes which are cylindrical or conical in shape.

The boundary conditions for $\Psi_m(r,z)$ on the outer FEM region and axis for both electrostatic and magnetic deflection problems are given as follows

$$\partial \Psi_m/\partial r = 0 \qquad \text{On axis} \qquad (10.11)$$

$$\Psi_m = 0 \qquad \text{On the outer boundaries} \qquad (10.12)$$

1.2   Element nodal equations

Minimising the functional U_m with respect to the nodal potentials at the ith node, Ψ_{mi}, provides the finite element nodal equations. The derivative with respective to the nodal potential Ψ_{mi} on element e that has area Ω_e is

$$\frac{\partial U_m}{\partial \Psi_{mi}} = \frac{\pi}{2} \left(\frac{\partial L}{\partial \Psi_m} + \frac{\partial R}{\partial \Psi_m} \right) \qquad (10.13)$$

where

$$
\frac{\partial L}{\partial \Psi_{mi}} = \iint_{\Omega_e} \gamma \left\{ 2r^{2m+1} \left(\frac{\partial \Psi_m}{\partial r} \frac{\partial}{\partial \Psi_{mi}} \frac{\partial \Psi_m}{\partial r} + \frac{\partial \Psi_m}{\partial z} \frac{\partial}{\partial \Psi_{mi}} \frac{\partial \Psi_m}{\partial z} \right) \right.
$$

$$
+ 2mr^{2m} \left(\Psi_m \frac{\partial}{\partial \Psi_{mi}} \frac{\partial \Psi_m}{\partial r} + \frac{\partial \Psi_m}{\partial \Psi_{mi}} \frac{\partial \Psi_m}{\partial r} \right)
$$

$$
\left. + 4m^2 r^{2m-1} \Psi_m \frac{\partial \Psi_m}{\partial \Psi_{mi}} \right\} drdz
$$

$$
\frac{\partial R}{\partial \Psi_{mi}} = 2gf_m \iint_{\Omega_e} \gamma \left\{ a_1 r^{m+1} \frac{\partial}{\partial \Psi_{mi}} \frac{\partial \Psi_m}{\partial r} + a_2 r^{m+1} \frac{\partial}{\partial \Psi_{mi}} \frac{\partial \Psi_m}{\partial z} \right.
$$

$$
\left. + \left(a_1 mr^m - a_3 mr^{m-1} \right) \frac{\partial \Psi_m}{\partial \Psi_{mi}} \right\} drdz
$$

(10.14)

The nodal potential Ψ_{mi} is expanded in terms of the shape functions $N_j(r,z)$

$$
\Psi_m = \sum_j N_j \Psi_{mj} \tag{10.15}
$$

so

$$
\frac{\partial L}{\partial \Psi_{mi}} = \iint_{\Omega_e} \gamma \left\{ 2r^{2m+1} \left(\frac{\partial N_i}{\partial r} \sum_j \Psi_{mj} \frac{\partial N_j}{\partial r} + \frac{\partial N_i}{\partial z} \sum_j \Psi_{mj} \frac{\partial N_i}{\partial z} \right) \right.
$$

$$
+ 2mr^{2m} \left(\frac{\partial N_i}{\partial r} \sum_j \Psi_{mj} N_j + N_i \sum_j \Psi_{mj} \frac{\partial N_j}{\partial r} \right) \tag{10.16}
$$

$$
\left. + 4m^2 r^{2m-1} N_i \sum_j \Psi_{mj} N_j \right\} drdz
$$

$$
\frac{\partial R}{\partial \Psi_{mi}} = 2gf_m \iint_{\Omega_e} \gamma \left\{ a_1 r^{m+1} \frac{\partial N_i}{\partial r} + a_2 r^{m+1} \frac{\partial N_i}{\partial z} + \left(a_1 mr^m - a_3 mr^{m-1} \right) N_i \right\} drdz
$$

(10.17)

The nodal equation at the ith node for n surrounding elements can then be expressed in the following form

$$\left[\sum_j K_{ij}\Psi_{mj} + G_i\right]_{\Omega_{e1}} + \left[\sum_j K_{ij}\Psi_{mj} + G_i\right]_{\Omega_{e2}} + \cdots\cdots\cdots \left[\sum_j K_{ij}\Psi_{mj} + G_i\right]_{\Omega_{en}} = 0$$

where

$$K_{ij} = \frac{\pi}{2}\iint_{\Omega_e}\gamma\left\{2r^{2m+1}\left(\frac{\partial N_i}{\partial r}\frac{\partial N_j}{\partial r} + \frac{\partial N_i}{\partial z}\frac{\partial N_i}{\partial z}\right) + 2mr^{2m}\left(\frac{\partial N_i}{\partial r}N_j + N_i\frac{\partial N_j}{\partial r}\right)\right.$$
$$\left. + 4m^2 r^{2m-1}N_i N_j\right\}drdz$$

$$G_i = \pi gf_m\iint_{\Omega_e}\gamma\left\{a_1 r^{m+1}\frac{\partial N_i}{\partial r} + a_2 r^{m+1}\frac{\partial N_i}{\partial z} + \left(a_1 mr^m - a_3 mr^{m-1}\right)N_i\right\}drdz$$

$$(10.18)$$

The global matrix is positive definite and symmetric.

2.   ACCURACY TESTS

A finite element program for deflectors was written based upon the method presented in the last section. Second-order 8 node serendipity elements are used and the program is compatible with the KEOS package [4]. To test the program and evaluate the accuracy of computed field solutions, three typical deflector configurations are analysed: cylindrical coils in free-space, infinitely long cylindrical coils with a magnetic core, and tapered coils in free space. The field distribution of the coils in free-space can be expressed analytically. In the case of a magnetic core, if the yoke is cylindrical and the ratio of its length to radius is large enough, a two-dimensional analytical solution can be used to calculate the fields in the middle of the deflection yoke. The accuracy of the program in treating the general case of a tapered deflector is also investigated. In some cases, the results of the finite element program is compared with a boundary element program written by Zhao [5].

2.1   The analytical solution for Magnetic Deflectors in free space

Analytical solutions for $d_1(z)$ and $d_3(z)$ can only be found in a few simple cases. For the toroidal and saddle coils in free space with a general tapered structure, as shown in Figure 10.1, the harmonic deflection functions $d_1(z)$ and $d_3(z)$ can be derived from Biot-Savart's law.

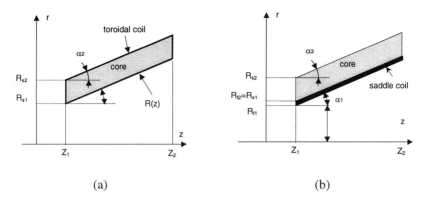

Figure 10.1 Schematic drawing of deflectors
(a) Toroidal yoke (b) Saddle yoke

For a toroidal coil with the geometrical parameters as defined in Figure 10.1a, a winding semi-angle of φ and ampere-turns of NI, the first and third harmonic deflection functions derived by Munro and Chu are as follows

$$
\left.
\begin{aligned}
d_1(z) &= \frac{NI}{\pi}\sin\varphi \int_{Z_1}^{Z_2}\int_{R_1(Z)}^{R_2(Z)} \frac{dRdZ}{L^3} \\[2em]
d_3(z) &= \frac{15\,NI}{8\pi}\sin 3\varphi \int_{Z_1}^{Z_2}\int_{R_1(Z)}^{R_2(Z)} \frac{R^2 dRdZ}{L^7}
\end{aligned}
\right\}
\tag{10.19}
$$

For saddle coils, the harmonic field expressions derived by Munro and Chu are only applicable to cylindrical saddle coils. The derivation for tapered saddle coils is given in Appendix 5. For the saddle coil depicted in Figure 10.1b, a winding semi-angle of φ, ampere-turns of NI, and distribution thickness ΔR ($=R_{t2}-R_{t1}$), the first and third harmonic deflection field functions are

$$
\left.
\begin{aligned}
d_1(z) &= \frac{NI}{\pi\Delta R}\sin\varphi \int_{Z_1}^{Z_2}\int_{R_1(Z)}^{R_2(Z)} \frac{R}{L^3}(\frac{3G}{L^2}-1)dRdZ \\[2em]
d_3(z) &= \frac{15\,NI}{8\pi\Delta R}\sin 3\varphi \int_{Z_1}^{Z_2}\int_{R_1(Z)}^{R_2(Z)} \frac{R^3}{L^7}(\frac{35G}{L^2}-\frac{5}{8})dRdZ
\end{aligned}
\right\}
\tag{10.20}
$$

where $L = \sqrt{R^2 + (z - Z)^2}$, $G = R^2 - R(z - Z)\tan(\alpha)$; $R_1(Z)$ and $R_2(Z)$ are the innermost and outermost equations respectively defining the radius of the winding profile between Z_1 and Z_2.

2.2   The analytical solution for a magnetic core deflection test example

To test the program accuracy for magnetic materials, a magnetic core deflector test example is considered. For this situation, Biot-Savart's law is no longer valid. For the simple case of an infinitely long cylindrical yoke, where a rotationally symmetric magnetic core has constant permeability, the problem becomes two-dimensional. In this case, the harmonic scalar potential distribution can be derived from Laplace's equation.

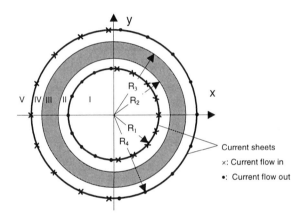

Figure 10.2 Two-dimensional model of a deflection yoke consisting of two current sheets of radii R_1 and R_4 on an angular magnetic core of radii R_2 and R_3.

The model used is shown in Figure 10.2 and consists of an angular magnetic core having a radius from R_2 to R_3 and two circular current sheets of radii R_1 and R_4 situated inside and outside the core. The relative permeability of the core μ_r is constant. The problem domain is subdivided into five current-free sub-domains, I, II, III, IV and V. In a cylindrical coordinate system (r, θ, z) in which the z-axis coincides with the yoke axis, Laplace's equation for the magnetic scalar potential Φ_m is

$$\frac{\partial^2 \Phi_m}{\partial r^2} + \frac{1}{r}\frac{\partial \Phi_m}{\partial r} - \frac{m^2 \Phi_m}{r^2} = 0 \qquad (10.21)$$

The angular current distribution on the radius R_1 is written as

$$i_{R_1}(\theta) = \sum_{m=1,3,\dots} i_m \cos m\theta \qquad (10.22)$$

To simulate a toroidal coil, the current on R_1 should return on R_4 so that the current forms a closed loop. This leads to

$$i_{R_4}(\theta) = \sum_{m=1,3,\dots} -i_m \cos m\theta \qquad (10.23)$$

The general form of the solution for Laplace's equation 10.21 is

$$\Phi_m(r) = C_m r^m + D_m r^{-m} \qquad (10.24)$$

This solution should satisfy the boundary conditions at the interfaces between the different sub-domains, i.e., it should have continuity of magnetic flux and satisfy Ampere's Law. Furthermore, it must remain finite throughout the domain. The solution in the deflection area (sub-domain I) is given by

$$C_m = \frac{i_m}{2mR_1^m}(1+\gamma_m) - \frac{i_m}{2mR_4^m}\lambda_m, \quad D_m = 0 \qquad (10.25)$$

Where γ_m is called the reflection factor of the core and is given by

$$\gamma_m = \frac{\left(\dfrac{R_1}{R_2}\right)^{2m}(\mu_r^2 - 1)\left[1 - \left(\dfrac{R_2}{R_3}\right)^{2m}\right]}{(\mu_r + 1)^2 - \left(\dfrac{R_2}{R_3}\right)^{2m}(\mu_r - 1)^2} \qquad (10.26)$$

and λ_m is called the transmission factor of the core and is given by

$$\lambda_m = \frac{4\mu_r}{(\mu_r + 1)^2 - \left(\dfrac{R_2}{R_3}\right)^{2m}(\mu_r - 1)^2} \qquad (10.27)$$

For saddle coils, the harmonic scalar potential can be obtained by making $R_4 \to \infty$ in the solution. The detailed derivation of all these expressions are presented by Osseyran [6].

2.3   FEM results for magnetic deflector test examples

An FEM model which can represent both a saddle and toroidal deflection yoke is illustrated in Figure 10.3. The dashed line indicates the outer boundary. The FEM simulation is performed separately for each coil type. Two typical configurations, i.e. cylinder ($\alpha_1 = \alpha_2 = 0°$) and tapered deflection yokes ($\alpha_1 = \alpha_2 \neq 0°$) are chosen for testing the programme. Their general parameters are listed in Table 10.1. An FEM region of 20 cm by 8 cm is used.

Winding parameters	Ampere-turns $=\pi$, Semi-Angle (radian) $= \pi/3$
Cylindrical yoke	$\alpha_1 = \alpha_2 = 0$, $L=1$, $R_{s1} = 0.45$, $R_{t1} = R_{s2} = 0.50$, $R_{t2} = 1.00$
Tapered yoke	$\tan(\alpha_1) = \tan(\alpha_2) = 1/2$, $L=2$, $R_{s1} = 0.45$, $R_{t1} = R_{s2} = 0.50$, $R_{t2} = 1.00$

Table 10.1 Cylindrical and tapered deflection yoke data (unit length: cm)

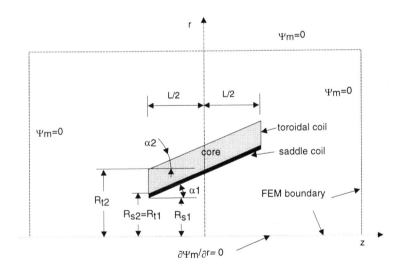

Figure 10.3 FEM model for saddle and toroidal yokes

An example of a typical mesh layout for the test deflector is shown in Figure 10.4. The shaded area corresponds to the core. A region block mesh layout is used and each region is automatically subdivided into a fine mesh. For the test examples presented here, a total of 2240 elements having 6919 free mesh nodes are used.

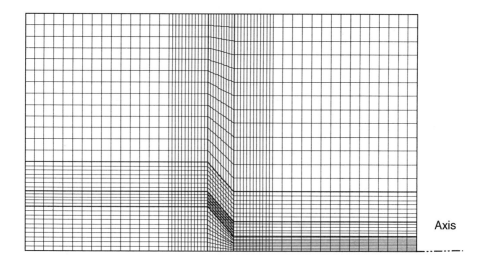

Axis

Figure 10.4 Typical mesh layout used in the air core magnetic deflector test
examples. The shaded region represents the core material

For the cylindrical yokes, the simulated deflection function $d_1(z)$ of toroidal and
saddle yokes are shown in Figures 10.5 (a) and (b) respectively. For the non-
magnetic core, the axial deflection $d_1(z)$ is calculated from Biot Savart's Law. The
simulated results are compared to analytical predictions. Their agreement is
excellent: 0.16% for the toroidal yoke and 0.10% for the saddle yoke at the peak
point.

Figure 10.5 shows that if the core is made of magnetic material, the curves of the
deflection function $d_1(z)$ tend to be flattened in the middle of the yoke as the
permeability increases. This implies that the field in the middle region is nearly two-
dimensional, and can be approximated by using the equation (10.24) given in the
previous section. The accuracy of this approximation is largely dependent on the
length, radius and permeability of the core. Table 10.2 gives both the numerical and
analytical prediction as well as percentage errors at the mid-point of the yoke for
different core permeabilities μ_r. For all these cylindrical magnetic coils, the results
agree well with the theoretical predications. The agreement for the air-core deflector
lies within 0.2 %. For high permeability, μ_r=10000, the agreement is typically within
0.05%. The higher discrepancy at low permeability for the toroidal coil yokes
(0.985% at μ_r=50) is due to inaccuracies of the two dimensional approximation. It
can be concluded that finite element errors for the relatively course mesh shown in
Figure 10.4 are typically less than 0.15%.

Type of deflector	μ_r	FEM (B/μ_0 AT)	Analytical (B/μ_0 AT)	Error (%)
Toroidal Yoke	1	163.155	163.422	0.163
	50	647.165	653.604	0.985
	100	669.139	672.641	0.521
	1000	690.250	690.741	0.071
	10000	692.500	692.613	0.016
Saddle Yoke	1	374.438	374.051	0.103
	50	673.248	672.925	0.048
	100	683.338	682.553	0.042
	1000	692.946	692.638	0.044
	10000	693.959	693.622	0.049

Table 10.2 Comparison of axial field results at the middle of a cylindrical deflector with a magnetic core

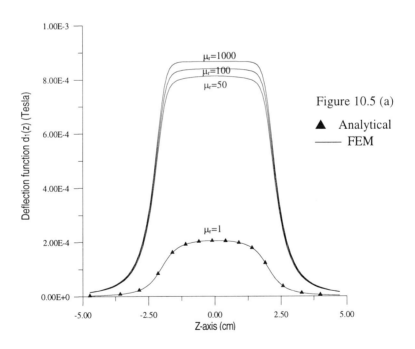

Figure 10.5 (a)

▲ Analytical

—— FEM

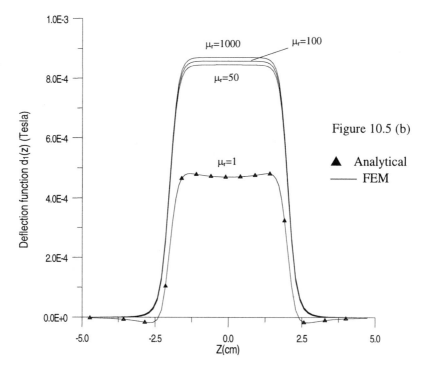

Figure 10.5 Simulation results for the deflection function $d_1(z)$ for an iron core of different magnetic permeabilities: (a) toroidal yoke (b) saddle yoke
L=4cm, Ampere-turns $=\pi$, Semi-Angle (radian) $= \pi/3$, $\alpha_1 = \alpha_2 = 0°$, $R_{s1} = 0.45$ cm,
$R_{t1} = R_{s2} = 0.50$cm, $R_{t2} = 1.00$ cm

Figure 10.6 shows the FEM axial field distribution $d_1(z)$ of air-core tapered saddle and toroidal yokes in comparison with their theoretical predictions. The results agree to within 0.064% and 0.23% for the saddle and toroidal yoke respectively at their core centres.   This high level of accuracy is similar to that obtained for the cylindrical air-core deflector test examples.

For a core made of magnetic material, the tapered deflector field distribution can only be solved by numerical methods. Figure 10.7 shows the comparison of the axial field distribution solved by the finite element and boundary element methods for tapered deflectors with an infinite permeability core. The results from the two numerical methods provide excellent agreement for the toroidal yoke (0.56%). The percentage error of 2.9% for the saddle yoke is a little higher, but this is mainly due to the different assumptions behind each numerical model. For instance, the thickness of the saddle coils are ignored in the boundary element model in order to treat it as a current sheet, while it is an important parameter for evaluating the current loading functions in the finite element program.

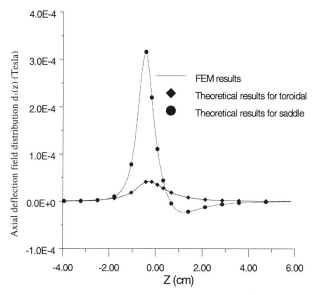

Figure 10.6 Axial field solutions of air-core tapered saddle and toroidal yokes obtained with FEM

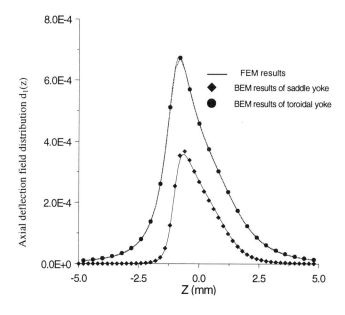

Figure 10.7 Axial field solutions of iron-core tapered saddle and toroidal yokes calculated by the FEM and BEM methods

The contours of the first harmonic scalar potential of the tapered saddle and toroidal yokes are illustrated in Figure 10.8. The contours for the saddle yoke are more concentrated around the central coil region than for the toroidal deflector, this is why the saddle yoke gives a steeper axial field distribution than the toroidal one.

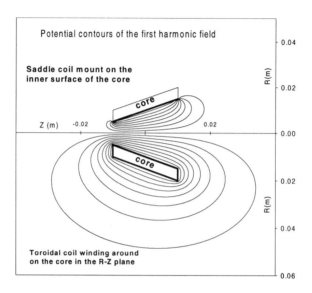

Figure 10.8 Contours of first harmonic scalar potential for saddle and toroidal yokes ($\mu_r = \infty$)

2.4   Electrostatic deflection

An octopole electrostatic deflector with cylindrical and conical profiles is used here as a test example. Figure 2.12 depicts its electrode layout. By defining an angular gap $\delta = \theta_{a1} = \pi/2 - \theta_{b2} = (\theta_{a2} - \theta_{b1})/2$ and $V_2 = \alpha V_1 = \alpha V$, the modified harmonic potential on electrodes Ψ_m on the cross-sectional plane at position z, given by equation (10.10), can be simplified to

$$\Psi_m(z) = \frac{4V}{\pi m^2 r(z)^m}\left\{(1-\alpha)\sin\frac{m\pi}{4}\sin m\delta + \alpha\cos(\frac{m\pi}{2} - m\delta)\right\} \qquad (10.28)$$

For the test example examined here, $\delta = 2.5\pi/180$ radians, V=1 and $\alpha = \sqrt{2} - 1$ radians. For this electrode voltage layout, it can be shown that both third-order and fifth-order harmonic terms are eliminated. The general arrangement and boundary conditions in the (r-z) plane are similar to those used for magnetic deflection. The coil geometry is defined by the parameters listed in Table 10.3.

Cylindrical arrangement	$\alpha_1 = \alpha_2 = 0$, L=1, R_{t1}= 0.50, R_{t2}=1.00
Conical arrangement	$\tan(\alpha_1) = \tan(\alpha_2) = 1/2$, L=2, R_{t1} = 0.50, R_{t2}=1.00

Table 10.3 Cylindrical and conical octopole electrostatic deflectors (units in cm)

To investigate the accuracy of the 2D harmonic FEM solution of the Octopole deflector, it is compared to the solution of a fully 3D FEM solution obtained from the commercial program TOSCA [7]. Since the third and fifth harmonics are not present in the test Octopole deflector, the 3D solution electric fields on the optic axis are compared directly to the first-order harmonic solutions. Only the first quadrant in the x-y plane is represented by the 3D mesh. The 3D mesh generated for the results presented here has 133500 linear elements and 135900 free nodes. The total node number in the transvers plane is around 4450. To give similar mesh conditions to the 3D mesh, the mesh for the 2D harmonic program consists of 1452 elements and 4511 free nodes in the (r,z) plane. Serendipity 8 node second-order elements are used. Figure 10.9 depicts first-order equi-harmonic-potential lines in the (r,z) plane for the conical shaped octopole electrostatic deflector. The harmonic potential Φ_1 on the surface of the electrodes is 1.0544 volts and it increases in steps of 0.09587 volts.

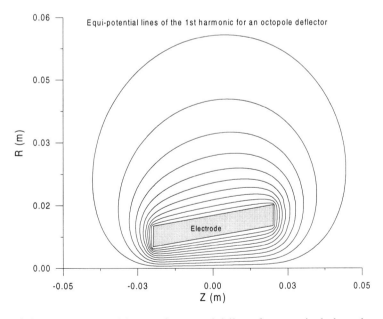

Figure 10.9 First-order equi-harmonic-potential lines for a conical shaped octopole electrostatic deflector

Figure 10.10 shows the FEM solved axial deflection fields for cylindrical and conical shaped octopole deflectors. The solid line represents the FEM results based on the 2D harmonic model, while the dotted points are generated from the fully 3D solution. The percentage agreement of the electric fields generated by these two methods for the modest mesh sizes used is less than 0.17% for the cylindrical arrangement and less than 0.3% for the conical arrangement. This high accuracy performance demonstrates the advantages of using the 2D harmonic FEM approach.

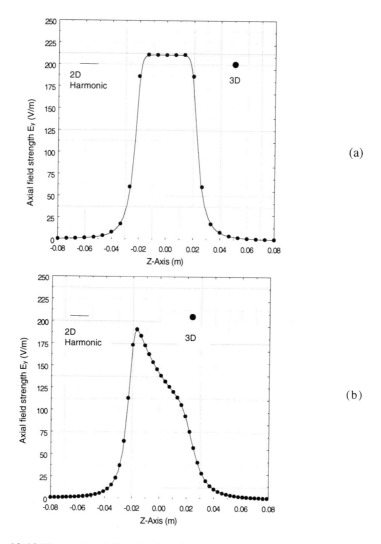

(a)

(b)

Figure 10.10 First-order deflection functions of octopole deflectors
(a) cylindrical (b) conical shapes

REFERENCES

[1]   E. Munro, and H.C Chu, "Numerical analysis of electron beam lithography systems. Part I: computation of fields in magnetic deflectors", Optik, 60, pp.371-390. 1982 and their paper, "Numerical Analysis of Electron beam systems. Part II: Computation of fields in electrostatic deflectors", Optik, 61,1-16, 1982

[2]   Lencová, B., Lenc, M. and van der Mast, K. D. Computation of magnetic deflectors for electron beam lithography. J. Vac. Technol. B, 7, pp.1846-1850, 1989.

[3]   Y. Zhao, "The design of In-Lens deflectors for electron beam systems", PHD thesis, Electrical Engineering Department, National University of Singapore, Singapore 119260, to be published

[4]   A. Khursheed, "The Khursheed Electron Optics Software" (KEOS), Department of Electrical Engineering, National University of Singapore, 1998

[5]   Y. Zhao, Numerical analysis of deflection systems in CRTs using a combined BEM and FEM method. M.Eng dissertation, Xian Jiaotong Uinversity, 1986

[6]   A. Osseyran, "Computer aided design of magnetic deflection systems", Phillips J. Res. 41, supplement 1, pp.34-36, 54-57. 1986

[7]   TOSCA, product of Vector Fields Ltd, 24 Bankside, Kidlington, Oxford OX5 1JE, Endgland, U.K.

Chapter 11
MESH RELATED ISSUES

The generation and refinement of finite element meshes has received much attention in recent years [1]. In charged particle optics, relatively little research has been done in this area. This section outlines some of the main issues linking the accuracy of finite element field solutions to mesh topology.

1.   STRUCTURED VS UNSTRUCTURED MESHES

The most common type of mesh generator used so far in electron optics is based upon creating region blocks [2]. Examples of this type of mesh have already been given in chapters 8 and 9. In two dimensions, the solution domain is typically divided into quadrilateral regions whose sides coincide with interfaces between different material types, such as iron/air interfaces in a magnetic circuit or conductor/air interfaces in an electrostatic structure. Each region is subdivided separately into a regular array of mesh lines that form the final numerical mesh. There are several disadvantages with this approach. The first is that the mesh density at each region boundary can change discontinuously and impose significant errors on the final potential distribution. Secondly, the specification of the regions is essentially a manual task, and for a problem having complex boundaries, it can be time-consuming. Thirdly, defining curved boundaries in terms of quadrilateral region blocks is cumbersome and inaccurate.

An important variation of the region block method is the graded mesh. Here the mesh spacing within a region is gradually increased or decreased, so that the change of the mesh spacing from cell to cell in any direction is small. The mesh spacing is made small for regions of high field intensity and large for regions of low field intensity. The change in mesh spacing between adjacent mesh cells is typically kept below 0.05, and is generated by a simple geometric algorithm. Kang et al [3] proposed an expanding mesh of this kind for electron gun simulation, while Lencová has advocated its use for the near axis region of electron lenses [4]. The graded mesh technique is obviously most effective for problems which have a single high field intensity region that is easily identifiable. An example of the graded mesh has already been presented in Chapters 8 and 9.

Another mesh type is based upon the Delaunay mesh generator [5,6]. This method is used both for mesh generation and mesh refinement. In two dimensions, it involves dividing the domain into different polygon regions and systematically subdividing the interior regions into triangle elements. Boundary points are specified as input data, and further additional points are added by trisection. Mesh nodes are connected in such a way so as to make the generated triangle elements as equilateral as possible.

The Delaunay mesh is refined by adding extra nodes according to a local error estimator of the field solution on an initial mesh. The density of the adaptive mesh typically varies with the local field strength. Unlike the region block mesh, the Delaunay mesh is usually unstructured. Delaunay mesh generation and adaptive refinement is used extensively in many areas of electromagnetics. Considerable research effort is presently being made to extend it to three dimensional field problems. Delaunay mesh generation has the merit of being more automated than the region block approach since it proceeds directly from boundary shape information.

A technique proposed by Khursheed is to refine the mesh according to equipotential and flux lines of the field solution [7] (see also chapter 9). The method is based upon constructing a curvilinear mesh from two conjugate potential solutions calculated on a trial mesh. The precise form of the trial mesh is not important since its main purpose is to generate field solutions from electrode structures that are topologically similar to the boundary shapes of the original problem. The mesh lines of the newly constructed mesh resemble the equipotential and flux lines of the final field solution and always cross one another at right angles. The density of the resulting mesh lines varies smoothly throughout the refinement domain. This type of mesh will be referred to here as a "quasi-curvilinear" refinement mesh.

Khursheed has also proposed a way of generating a structured mesh automatically from boundary shape information, without the need to create regions. The technique is based upon moving the nodes of a background regular mesh onto electrode boundaries and then using the boundary-fitted coordinate method to adjust all other mesh node positions automatically [8]. This type of mesh generation will be described in the next section.

Figure 11.1a shows the schematic of a single pole magnetic immersion lens similar to the one discussed by Shao and Lim [9]. High accuracy field computations are required in the simulation of this lens both for the primary and secondary electrons. Firstly, in order calculate the aberrations imposed upon the primary beam by this lens, the first few derivatives of the axial field distribution must be continuous, and this requires a high degree of accuracy in the axial field computation [10]. On the other hand, since the secondary electrons travel back up through the lens, the influence of stray magnetic fields on the secondary trajectory paths needs to be investigated, which requires the accurate computation of the field distribution in the lens bore.

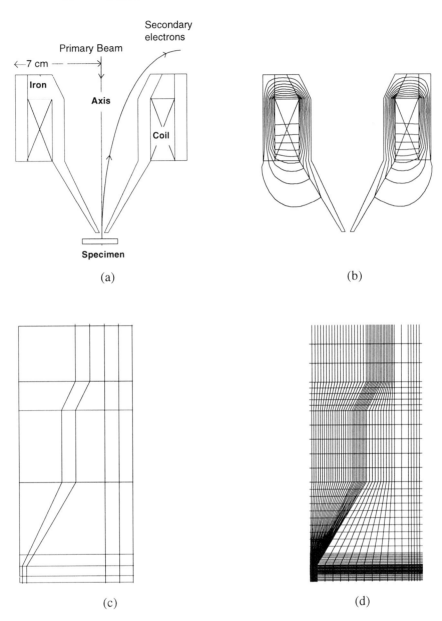

Figure 11.1 Region block model of a single pole immersion lens
    (a) Lens schematic
    (b) Flux lines from the KEOS program
    (c) Region lines
    (d) Mesh lines after region subdivision

Figure 11.1b shows flux lines for the single pole immersion lens produced by KEOS, the Khursheed Electron Optics Software [11]. This suite of finite element programs uses a region block mesh method to calculate electrostatic and magnetostatic field distributions in two and three dimensions. The region block mesh and its subsequent subdivision is shown in Figures 11.1c and 11.1d. The final mesh is composed of 48 by 63 lines. The region lines are chosen to fit the magnetic circuit, and the subdivision allocates a greater mesh density to the single pole piece tip region, where the axial field strength is expected to be highest. The bottom edge of the domain is specified to be a natural boundary.

A Delaunay mesh layout and its subsequent refinement was made for the single pole immersion lens by using OPERA2D, a commercial finite element program [12]. Figures 11.2a-d depict mesh lines for the unstructured initial mesh and its adaptive mesh refinement. Four iterations of adaptive mesh refinement increased the 1431 nodes of the initial mesh to 4652 nodes in the final mesh. The initial number of 2720 elements was increased to 9162. As shown by Figures 11.2a-d, the extra nodes added to the initial mesh are situated largely at the iron/air interface.

A curvilinear refinement mesh was created by plotting equipotentials on the two conjugate capacitor field problems shown in Figures 3a and 3b. Both electrode structures are similar in form to the magnetic circuit of the single pole immersion lens. For the first capacitor layout, the axis is specified to be at 0V, while the electrode corresponding to the iron part of the magnetic circuit is assigned to 1V. Natural boundaries lie at the top and bottom of the domain. The boundary conditions for the second capacitor layout are found by interchanging the fixed boundary conditions with the natural ones in the first capacitor layout. Both the potential distributions for the electrode structures shown in Figures 11.3a and 11.3b were calculated in two dimensional rectilinear coordinates. The equipotential lines of each solution are overlaid on top of one another and they form the refined mesh. Intersections between the two sets of equipotential lines can easily be calculated by linear interpolation. The choice of a rectilinear coordinate system is necessary to produce a conformal refinement mesh, that is, one where each mesh cell is a curvilinear square.

An example of the conformal refinement mesh generated is shown in Figures 11.3c and 11.3d. These figures demonstrate that the mesh density is highest in the vicinity of the single pole tip and that it varies smoothly throughout the refinement domain. The refinement mesh in this case is composed of 12 by 100 lines. The magnetic field distribution was re-solved on the refinement mesh. The boundary conditions for the vector potential at the domain edges were found by using linear interpolation on the initial region block mesh field solution.

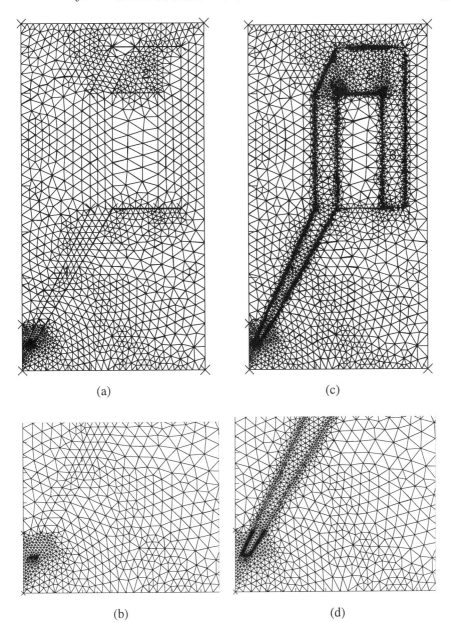

(a)   (c)

(b)   (d)

Figure 11.2 Unstructured mesh generation and refinement for the single pole immersion lens
(a) Initial mesh for the whole domain (b) Initial mesh around the pole tip
(c) Refined mesh for the whole domain (d) Refined mesh around the pole tip

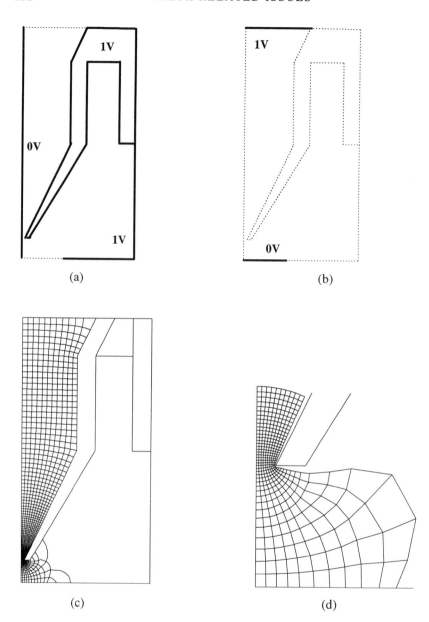

Figure 11.3 Structured refinement for the single pole immersion lens
(a) First capacitor layout
(b) Second capacitor layout
(c) Mesh lines constructed from equipotential lines on the conjugate capacitor
      solutions in rectilinear coordinates
(d) Magnified view of the mesh lines around the pole tip

Figure 11.4 shows that the solution on the axial field distribution from the Delaunay refined mesh is not significantly different to the one produced by the initial unstructured mesh. The reason for this is apparent from Figures 11.2a-d. They show that while the total number of elements is increased by a factor of three, the mesh density around the optic axis is not significantly changed. This comes from the global nature of the local error estimator, which finds the elements located at the iron/air interface to have the greatest error, and therefore refines them first. In electron optics, where the accuracy of the field solution on and around the optic axis is the most important concern, this kind of automatic refinement strategy is clearly not suitable.

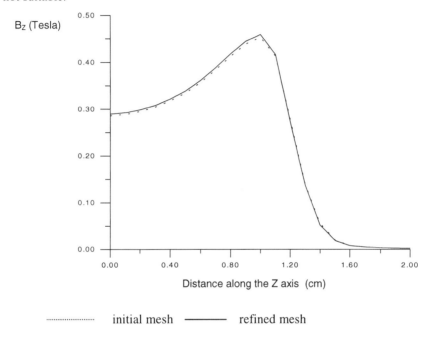

initial mesh ─────── refined mesh

Figure 11.4 Axial magnetic field distribution for the unstructured mesh

Figures 11.5a and 11.5b depict the first and second derivatives of the axial field distribution respectively. These figures show that the region block and Delaunay meshes produce large discontinuities in the axial field derivatives, which are generated by mesh density discontinuities. The conformal refinement mesh in comparison, generates smooth variations in the axial field distribution and is continuous up to the third derivative. These results clearly show the advantage of using structured refinement. Similar results have already been presented in the context of analysing the axial field continuity of magnetic lenses in chapter 9, where it was shown that graded meshes also provide a way of generating continuous axial field derivatives. Based upon these kinds of observations, it can be concluded that structured meshes are better suited for the field distributions of electron optics than unstructured ones.

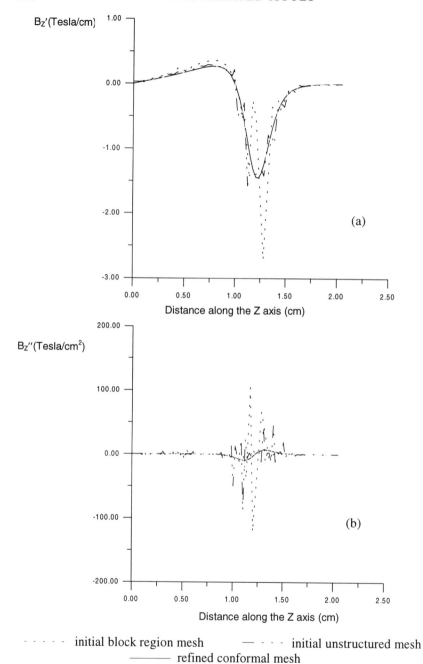

Figure 11.5 Derivatives of the axial field distribution for the single pole lens
(a) First derivative   (b) Second derivative

2.   THE BOUNDARY-FITTED COORDINATE METHOD

The boundary-fitted coordinate method has been extensively used in the field of aerodynamics and fluid flow analysis [13]. In the field of electron optics, Tagaki et al have used it to help solve for the space charge distribution of non-cylindrical electron guns [14]. But unlike this previous work, which for the most part applied the boundary-fitted coordinate method to finite difference field solving schemes, the following section highlights the advantages of using it as a mesh generator for finite element methods. It will be shown that the boundary-fitted coordinate method is a flexible way of modelling curved boundaries. It can also be used in automatic mesh generation, where only boundary shape data is required, avoiding the use of region blocks. It has these desirable features while at the same time retaining the advantages of a structured mesh.

One way of understanding the boundary-fitted coordinate method is to view it as a numerical means of mapping an irregular mesh defined in real space onto a logical mesh in normalised space. In two dimensions, this corresponds to a procedure of mapping a region R(x,y) onto a region in normalised space, N(u,v), where u and v are the coordinates in normalised spaced, as shown in Figure 11.6. The normalised coordinates (u,v) appear as functions in Poisson's equation

$$\nabla^2 u = f_1(u, v) \qquad \nabla^2 v = f_2(u, v) \qquad\qquad (11.1)$$

where the Laplacian is defined in terms of the usual (x,y) coordinates and the functions f_1 and f_2 are specified in terms of the normalised coordinates. These functions allow for greater flexibility in the mesh generation process and can for instance, be used to vary the mesh density locally in real space.

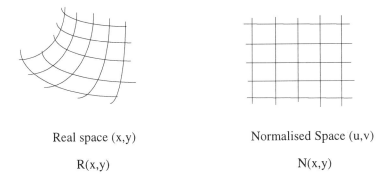

Real space (x,y)   Normalised Space (u,v)

R(x,y)   N(x,y)

Figure 11.6 The Boundary-Fitted Coordinate method mapping

As with many other mapping techniques for solving field problems, the transformation is defined in normalised space and it is the inverse problem that needs to be solved. In terms of the coordinates at the mesh nodes in real space x(u,v) and y(u,v), and their single and double derivatives in the u-v plane, the boundary-fitted coordinate method involves solving the following pair of coupled equations [15]

$$ax_{uu} + 2bx_{uv} + cx_{vv} + \det(J)^2[Px_u + Qx_v] = 0$$
$$ay_{uu} + 2by_{uv} + cy_{vv} + \det(J)^2[Py_u + Qy_v] = 0 \qquad (11.2)$$

where J is the Jacobian of the transformation, $P(u,v)$ and $Q(u,v)$ are functions that control the local mesh spacing. J, a, b, and c are given by

$$a = x_v^2 + y_v^2 \quad b = -(x_u x_v + y_u y_v) \quad c = x_u^2 + y_u^2 \quad \det(J) = x_u y_v - x_v y_u \qquad (11.3)$$

In three dimensions, the coordinates of the transformed point (u,v,w) are solutions to the following set of coupled equations

$$\nabla^2 u = f_1(u, v, w) \qquad \nabla^2 v = f_2(u, v, w) \qquad \nabla^2 w = f_3(u, v, w) \qquad (11.4)$$

Let the matrix M be defined by

$$M = \begin{bmatrix} x_u & x_v & x_w \\ y_u & y_v & y_w \\ z_u & z_v & z_w \end{bmatrix} \qquad (11.5)$$

and let J be the determinant of matrix M. Given the boundary conditions of the electrodes in normalised space, it can be shown that the inverse problem of finding $x(u,v,w)$, $y(u,v,w)$ and $z(u,v,w)$ involves solving the following set of coupled equations [16]

$$\alpha_{11}x_{uu} + 2\alpha_{12}x_{uv} + 2\alpha_{13}x_{uw} + \alpha_{22}x_{vv} + 2\alpha_{23}x_{vw} + \alpha_{33}x_{ww} + J^2[f_1 x_u + f_2 x_v + f_3 x_w] = 0$$
$$\alpha_{11}y_{uu} + 2\alpha_{12}y_{uv} + 2\alpha_{13}y_{uw} + \alpha_{22}y_{vv} + 2\alpha_{23}y_{vw} + \alpha_{33}y_{ww} + J^2[f_1 y_u + f_2 y_v + f_3 y_w] = 0$$
$$\alpha_{11}z_{uu} + 2\alpha_{12}z_{uv} + 2\alpha_{13}z_{uw} + \alpha_{22}z_{vv} + 2\alpha_{23}z_{vw} + \alpha_{33}z_{ww} + J^2[f_1 z_u + f_2 z_v + f_3 z_w] = 0$$

where

$$\alpha_{jk} = \sum_{m=1}^{N} \beta_{mj}\beta_{mk} \qquad (11.6)$$

and β_{jk} is the cofactor of the (j,k) element in the matrix M.

The (x,y,z) coordinates of the nodes on the boundary surfaces are fixed, and the (x,y,z) coordinates of the non-boundary nodes can be found by using the finite difference method in combination with the successive over-relaxation scheme. For the purpose of generating a finite element mesh, high accuracy in the mesh node positions is not required and only a small number of iterations are adequate. In practice, around 10-30 iterations was found to be sufficient. Beyond this number, the free node mesh coordinates do not change significantly.

It should be noted that the electric potential $\Phi(u,v,w)$ also obeys the same equation as the coordinates $x(u,v,w)$, $y(u,v,w)$ and $z(u,v,w)$, that is

$$\alpha_{11}\Phi_{uu} + 2\alpha_{12}\Phi_{uv} + 2\alpha_{13}\Phi_{uw} + \alpha_{22}\Phi_{vv} + 2\alpha_{23}\Phi_{vw} + \alpha_{33}\Phi_{ww} + J^2\left[f_1 x_u + f_2 x_v + f_3 x_w\right] = 0$$

$$(11.7)$$

This means that for electric field problems which do not have any dielectric materials, and where all curved boundaries require Dirichlet conditions, the electric potential distribution can be directly solved by the finite difference method. But this is not recommended, since the truncation noise at the mesh nodes resulting from a finite-difference approximation to the above equation is much greater than solving Laplace's equation in real space. Also, for the more general case where there are dielectric materials or where Neumann boundary conditions need to be applied on curved boundaries, the finite element method is more flexible and the boundary-fitted coordinate technique is used as a means to generate the numerical mesh.

The simplest way to generate a boundary-fitted mesh is to map single regions at a time, where the nodes on each region $R(x,y)$ in real space have a one-to-one correspondence to the nodes on the sides of the square region $N(u,v)$ in normalised space. Obviously this procedure is suitable for the outer boundaries of the domain, as shown in Figure 11.7a. Here the outer boundaries of the domain are mapped onto a circle and the coordinates of the internal mesh nodes are solved by the boundary-fitted coordinate method. Conductors can then be laid out onto the new mesh, as in this case, where a quadupole structure is specified. The equipotential lines of the field solution are shown in Figure 11.7b.

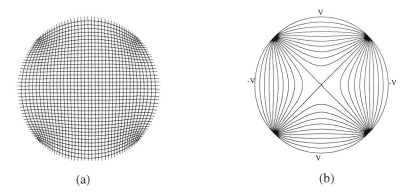

(a)   (b)

Figure 11.7 Single region boundary-fitted coordinate mapping for an electrostatic quadrupole   (a) Mesh   (b) Equipotential lines

A single domain application of the boundary-fitted coordinate method can also be applied to generate the near-axis region of an electron gun. This is different from Tagaki et al who use the boundary-fitted coordinate method and domain decomposition to create multiple regions. The mesh distribution on the region boundaries is defined by the well-known parabolic conformal mapping. The potentials at nodes on the right-hand side and top boundaries are obtained by interpolation from a background potential distribution. The mesh used for the background potential distribution uses a coarse mesh and covers the entire gun layout. Figure 11.8 shows part of a single domain boundary-fitted coordinate mesh which has been generated around the cathode of a field emitter gun.

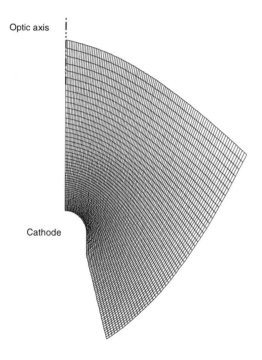

Figure 11.8 Single region boundary-fitted coordinate method for generating the near-cathode region mesh of a field emission electron gun

It is also possible to map rectangular conductor region blocks in the normalised domain to curved boundary blocks in real space. The coordinates at the edges of each block are first mapped and fixed, then the coordinates of all the other mesh nodes are solved by applying the boundary-fitted coordinate method. Figure 11.9 shows an example of how two square blocks in the normalised domain are transformed to circles in real space. However, the corner mesh cells in normalised space can appear quite distorted in real space, and this way of creating curved boundaries is not the optimal way of using the boundary-fitted coordinate method.

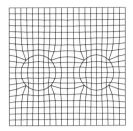

Figure 11.9 Boundary-fitted coordinate mesh to fit two circular electrodes based upon mapping of individual conductor blocks

The method of mesh generation recommended here is to start with a logical background structured mesh defined in Cartesian coordinates. Intersections of the mesh lines with electrode boundaries are first calculated and stored. Then the positions of those nodes lying nearest to electrode boundaries are moved onto the electrode surface. The mesh now is locally distorted to accommodate the presence of electrode boundaries and the boundary-fitted coordinate method is used to adjust the position of non-boundary nodes so that they are distributed smoothly throughout the domain. The main advantage of this approach is that it minimises mesh distortions and that it can be automated. It requires only boundary information to generate the mesh and obviates the need to specify regions.

Figure 11.10 shows an example of mesh generation for a coaxial cable layout along with its field solution. Note that extra triangle elements may be generated at the boundaries. Figure 11.11 shows the mesh generation and field solution of deflector plates whose boundaries are composed of curved and straight parts. The boundary information is specified here in terms of the coordinates of polygons and arcs of circles. Clearly the boundary-fitted coordinate method has the flexibility of automatically generating a structured mesh directly from boundary shape information.

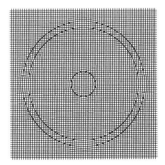
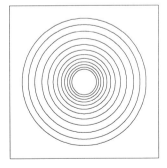

Figure 11.10 Boundary-fitted coordinate mesh for coaxial layout based upon minimal mesh node movement

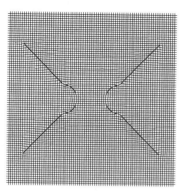 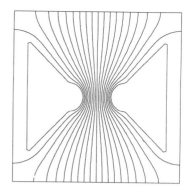

Figure 11.11 Boundary-fitted coordinate mesh for an electrostatic deflector based upon minimal mesh node movement

The minimal mesh perturbation technique can of course be generalised to three dimensions. A schematic of a spherical capacitor is given in Figure 11.12a. The inner electrode has a radius of 2 cm and carries a potential of 1 V. The outer electrode has a radius of 8 cm and is specified to be at 0 V. A mesh resolution of 21 by 21 by 21 lines is used, which results in 1796 free nodes in the region between the electrodes. The mid-plane mesh through the spherical electrodes lies at IZ=11. The plan view of the mid-plane and the equipotential lines on its surface are shown in Figures 11.12b and 11.12c. FEM programs using first-order tetrahedral elements are used to calculate the potential distribution. The root mean square error on the node potentials was found to be 4.3×10^{-3} volts. Figures 11.12d-h show a perspective view of the X-Y mesh planes as IZ varies from 3 to 11. These figures show how mesh nodes are locally perturbed to fit the curved surface of the electrodes and demonstrate that the mesh nodes are distributed smoothly throughout the domain.

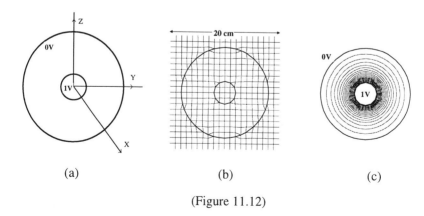

(a)   (b)   (c)

(Figure 11.12)

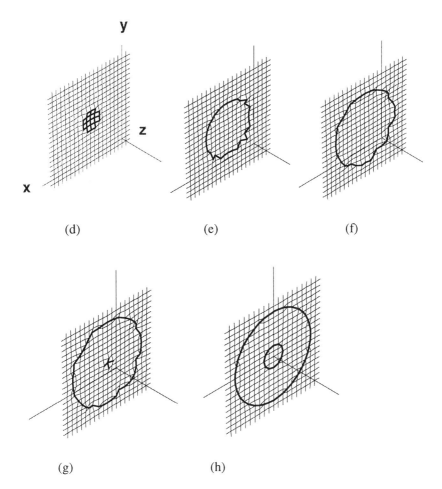

Figure 11.12 Finite element solution to the spherical capacitor structure on a 21 by 21 by 21 boundary-fitted coordinate generated mesh defined in rectilinear coordinates
(a) Electrode schematic (b) Plan view of mesh at the mid-plane (IZ=11)
(c) Equipotential lines on the mid-plane   (d) Mesh plane at the outer electrode (IZ=3) (e) Mesh plane for IZ=5 (f) Mesh plane for IZ=7
(g) Mesh plane for IZ=9 (h) Mesh plane for IZ=11 (mid-plane)

The method of moving nodes of a structured mesh onto curved boundaries and using the boundary-fitted coordinate method to redistribute nodes smoothly is especially useful for curved material boundaries. Here, the method described in section 2 Chapter 6, of automatically subdividing brick elements into tetrahedra that fit the curved interface is used. Results for the spherical capacitor example with a dielectric layer around the inner conductor are shown in Figure 11.13. A mesh of 41 by 41 by

41 lines is used and the dielectric layer has a relative permittivity of 5. The accuracy of the nodal potentials were calculated to be less than 0.1%.

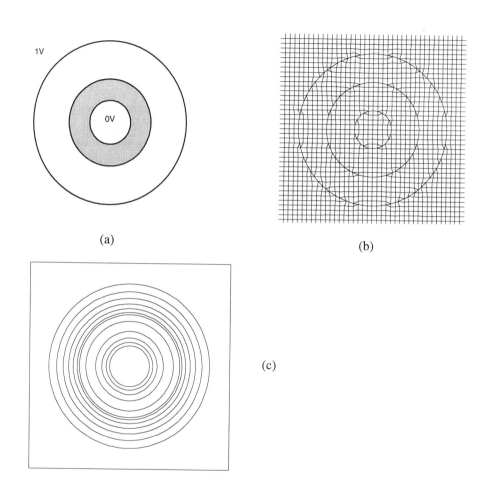

(a)   (b)

(c)

Figure 11.13 Spherical Capacitor with dielectric layer (ε_r=5)
(a) Layout (b) Mesh (c) Equipotentials

3.   MESH REFINEMENT FOR ELECTRON GUN SIMULATION

In the numerical simulation of electron guns, fast accurate computations of electron trajectories are required. The space charge distribution around the cathode is calculated by solving a self-consistent problem involving the electric field distribution around the cathode and electron trajectories that are emitted from it. An iteration procedure is often used, where the current extracted by the electric field distribution is estimated by a simple one-dimensional solution to Poisson's equation such as that described by Child's Law. The space charge distribution deposited by electrons around the cathode can then be calculated by plotting their trajectory paths. The effect of the space charge distribution on the electric field distribution is then calculated from a numerical solution to Poisson's equation, and more electron trajectories are plotted. The process is repeated until the current extracted from the cathode stabilises [17].

Using the finite element method to solve Poisson's equation for the field distribution around the electron gun presents several difficulties. The first problem is that the numerical mesh must model boundary shape variations that can vary by several orders of magnitude. In a cold field emission gun for instance, the radius of the cathode tip is typically 0.1 microns, approximately 0.0001 times smaller than the cathode-anode distance. In thermal triode guns, the cathode tip radius is typically 0.001 times smaller than the cathode-anode distance. The mesh spacing around the cathode tip should be several times smaller than the cathode tip radius.

A quasi-curvilinear mesh for a wide variety of different electron gun layouts is presented here. This technique was first reported by Khursheed for electron guns and was later generalised to electron lenses [18]. The block region method for generating the mesh is used to solve trial potential problems. Equipotential lines on the trial potential distribution are used to construct the new mesh. Due to the regular structure of the quasi-curvilinear mesh, a Hermite-Cubic/Spline interpolation method can be used to derive accurate electric values at any point in the domain. Spline interpolation can be used at the start of the program to pre-compute derivatives and cross-derivatives of the potential and mesh coordinates at every mesh node in the domain. Subsequent field evaluations at any point in the domain through Hermite-Cubic interpolation uses only a function's values, derivatives and cross-derivatives at four local mesh nodes. The plotting of electron trajectories can thus be greatly improved, both in terms of computation speed and accuracy.

Figure 11.14a shows the electrode layout for a thermionic electron gun and Figure 11.14b shows the region block division used to create a trial mesh for it. The region blocks were linearly subdivided, and a total of 80 by 100 mesh nodes were used for the fine mesh. Figures 11.14c and 11.14d show how this electrode layout can be arranged into two conjugate capacitor field problems. The boundary conditions are such that the fixed and natural boundaries of the first capacitor are interchanged to derive the boundary conditions for the second capacitor field problem. Once each field problem has been solved using the trial mesh, equipotential lines are plotted and overlaid on top of one another. Figure 11.14e shows a quasi-conformal mesh of 16 by 96 mesh lines produced by this method where the conjugate capacitor field distributions were solved in rectilinear coordinates. This method of mesh generation is obviously a convenient way of concentrating mesh points around the cathode where the electric field strength is highest. The mesh spacing above the cathode tip is 1.66×10^{-4} times smaller than the cathode-anode distance. This mesh can be directly used to model the cathode tips used in thermionic triode guns.

Meshes can also be generated by trial potential distributions that are solved in axisymmetric cylindrical coordinates (r,z). Figures 11.14f shows a curvilinear mesh generated for the thermionic electron gun. The mesh resolution is 16 by 96, and the mesh spacing directly above the cathode tip is 2×10^{-5} times smaller then the distance separating the cathode from the anode. This mesh resolution is sufficiently high enough to directly model field emission tips.

Figures 11.14e and 11.14f show meshes that can be used to model the whole domain as well as the cathode tip directly and are suitable for electron gun simulation. It must be noted that an accurate representation of the cathode tip must first be modelled by the trial mesh. If the region block mesh generator is used, it should be able to concentrate many mesh points around the cathode tip. A Delaunay triangle mesh can also be employed. Once a trial mesh has been generated, then the equipotential line method for generating a quasi-curvilinear mesh can be used to produce a single optimal mesh for the whole domain.

It should also be noted that equipotential lines need not be plot for constant potential intervals. Indeed, one way of significantly increasing the mesh density where required is to plot equipotentials at finer potential steps. This procedure is a convenient way of varying the mesh density throughout the domain.

A near-cathode quasi-curvilinear mesh, similar to the one shown in Figure 11.8 generated by the boundary-fitted coordinate method can also be created by solving two conjugate field problems. A quasi-conformal mesh for the near-cathode region is shown in Figure 11.15. The two conjugate capacitor potential distributions provide equipotential lines that start from the cathode and finish on the top boundary, and a set of lines that start from the optic axis finish on the right-hand side boundary. The mesh around the cathode tip is better than the one created by the boundary-fitted coordinate method. This is because mesh lines intersect the cathode at right-angles, resulting in a lower truncation noise and simplifying the calculation for the electric field components close to the cathode surface. Accurate field values at the cathode surface are not only important for trajectory plotting, but in the case of field emission guns, are also required to determine the emitted current. An example in the next section will demonstrate that this type of quasi-curvilinear mesh around the cathode provides very accurate field values.

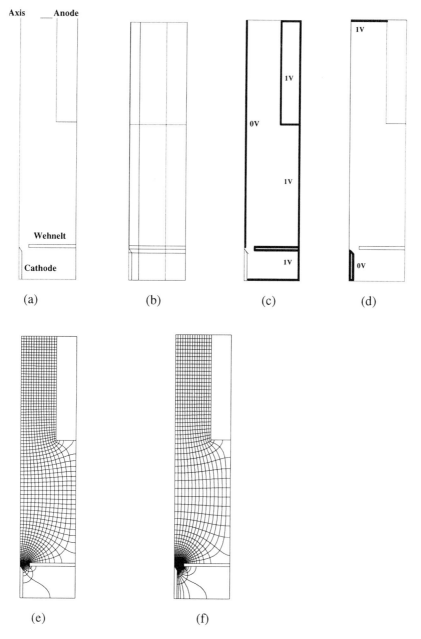

Figure 11.14 Thermionic Gun model
(a) Layout (b) Region Block representation (c) Fixed left-hand and right-side
(d) Fixed top and bottom (e) Mesh from solution in (x,y) (f) Mesh from
solution in (r,z)

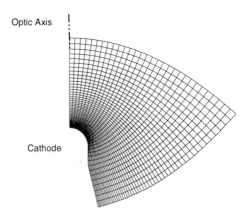

Figure 11.15 Quasi-curvilinear mesh for a near-cathode region of a field emission electron gun

Trajectory plotting can be performed in the normalised (u,v) plane. This greatly simplifies the trajectory plotting procedure, since accurate electric field values can be obtained directly by Bi-Cubic Spline interpolation in the u-v plane. The equations of motion in the longitudinal direction z, and radial direction r are given by the following,

$$\ddot{r} = -\eta \phi_r$$

$$\ddot{z} = -\eta \phi_z$$

(11.8)

where ϕ_r and ϕ_z are spatial derivatives of the electric potential $\phi(r,z)$, and η is the electron charge to mass ratio. These equations can be transformed to the u-v plane, giving the following pair of coupled equations

$$a\ddot{u} + b(\dot{u})^2 + c\dot{u}\dot{v} + d(\dot{v})^2 = e$$

$$f\ddot{v} + g(\dot{v})^2 + h\dot{u}\dot{v} + i(\dot{u})^2 = k$$

(11.9)

where

$$a = r_u z_v - z_u r_v \quad b = r_{uu} z_v - z_{uu} r_v \quad c = 2(r_{uv} z_v - z_{uv} r_v) \quad d = r_{vv} z_v - z_{vv} r_v$$

$$f = r_v z_u - z_v r_u \quad g = r_{vv} z_u - z_{vv} r_u \quad h = 2(r_{vu} z_u - z_{vu} r_u) \quad i = r_{uu} z_u - z_{uu} r_u$$

$$e = \eta z_v (c_{11}\phi_u + c_{12}\phi_v) - \eta r_v (c_{21}\phi_u + c_{22}\phi_v) \quad k = \eta z_u (c_{11}\phi_u + c_{12}\phi_v) - \eta r_u (c_{21}\phi_u + c_{22}\phi_v)$$

$$\det J = r_u z_v - r_v z_u \quad c_{11} = \frac{z_v}{\det J} \quad c_{12} = \frac{-z_u}{\det J} \quad c_{21} = \frac{-r_v}{\det J} \quad c_{22} = \frac{r_u}{\det J}$$

(11.10)

The derivatives and cross-derivatives of r(u,v), z(u,v) are found by performing Bi-Cubic Spline interpolation in the normalised plane, taking the values of r and z at each mesh node. Just as with the electric potential $\phi(u,v)$, derivatives and cross-derivatives of r and z at the mesh nodes are calculated at the beginning of the program and stored for subsequent use in a Hermite-Cubic interpolation routine.

The total time to plot an electron trajectory path by this method is comparable to using linear interpolation to derive electric field values on an unstructured mesh. This is because locating the electron's position with respect to the mesh nodes on an unstructured mesh is a complicated procedure, where the shape functions for each element are used to detect whether the electron falls outside or inside an element. The method of plotting the electron trajectory directly in normalised space avoids this problem, since locating the electron's position in normalised coordinates with respect to the mesh lines is a trivial task.

Figures 11.16a and 11.16b show electron trajectories that are traced from the cathode of a field emission gun. In Figure 11.6a, electrons leave the cathode with zero initial energy and are traced around the cathode surface. There is one trajectory per mesh cell and the maximum semi-angle for the trajectories is 0.1 radians. Figure 11.6b shows electron trajectories that leave the centre of the cathode with initial energies of 1 eV. These electrons are emitted over a wide range of initial angles with an angular step size of $\pi/10$ radians. The electron gun layout for which these trajectories were plotted is shown in Figure 2.1, and part of the quasi-conformal mesh near the cathode is depicted in Figure 11.15. The potential distribution on the outer boundaries of the quasi-conformal mesh were found by interpolation from a background potential distribution which spans the whole domain. A region block mesh was used for the background potential distribution.

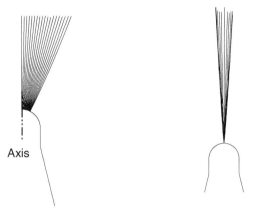

Figure 11.16 Electron trajectory paths leaving the cathode of a field emission gun
(a) Zero energy electrons plot up to a final sem-angle of 0.1 radians
(b) 1 eV energy electrons plot in angular steps of $\pi/10$ radians

4.   HIGH-ORDER INTERPOLATION

Since it is either electric or magnetic potentials which are output at mesh nodes by finite element programs, field information at any point in the domain requires the use of interpolation and numerical differentiation. One obvious way of providing the spatial derivatives of the potential distribution on the mesh is to use finite element shape functions. These functions usually take the form of low order polynomial expansions over the element. Low order here is taken to mean second-order and lower. Use of linear function variations over a triangle element is particularly widespread. This is because such an element can readily be incorporated into a wide variety of different mesh generation and refinement techniques. The method of using the first-order shape function for post-processing purposes will be referred to here as "linear interpolation".

High-order methods of interpolation are typically two to four orders of magnitude more accurate than linear interpolation. For curvilinear mesh cells, such as finite-difference meshes, the advantages of high-order interpolation are well known [19]. Interpolation based upon finite element shape functions leads to large discontinuities of the field values at the element edges. For these reasons, where the best available accuracy for field computation is required, as is often the case in electron optics, high-order methods of interpolation are recommended.

It is also important to note that the methods presented here are quite different to finite element derivative computations presented elsewhere, such as those used for force computations [20] or those methods which have been devised to improve the solution of the inverse problem [21]. In these situations, derivatives were found by directly differentiating the final finite element global matrix equation and then solving for the derivatives as part of a new matrix equation. But here it is the derivative of the potential distribution that is sought, which is quite different to minimising the virtual work of a system, or optimising the system with respect to certain design parameters. It might be suggested that the procedure of directly differentiating the finite element global matrix equation be applied to the matrix equation for the potential distribution. But a little further consideration shows that this procedure will only result in an indirect method of using the finite element shape functions, and hence will produce results comparable in accuracy to those based upon using the shape functions directly.

Consider a stiffness matrix [A], a matrix [B] for the source terms, and a potential distribution denoted by [V], so that the final matrix equation of the finite element solution for [V] is given by [A][V]=[B]. Taking the derivative of this equation with respect to the x coordinate leads to the following equation

$$[A]\frac{d[V]}{dx} = \frac{d[B]}{dx} - \frac{d[A]}{dx}[V] \qquad (11.11)$$

Here it might be argued that by solving this equation and a similar derivative equation with respect to the y coordinate, the potential derivatives can be found directly. But what is found by solving these matrix equations is the potential derivatives at the mesh nodes only. For applications such as electron trajectory plotting, field values are required at all points along the trajectory path. Clearly another step of interpolation is necessary and it must be critically evaluated for its level of accuracy. More importantly, the derivatives of the matrices [A] and [B] are evaluated in terms of the finite element shape functions, so this method is limited by the order and continuity constraints of the shape functions used.

Devising high-order methods of interpolation on finite element meshes is a non-trivial task. This difficulty has been cited to be a major weakness of the finite element in electron optics. Hawkes and Kasper claim that accurate computation of the field strength on FEM meshes is "practically impossible" [22]. The work presented here shows that this statement is incorrect. In fact, Khursheed has been using high-order interpolation on finite element meshes to simulate the trajectory paths of secondary electrons in scanning electron microscopes since the early eighties. The basis of this technique was published in 1989 [23]. Special techniques need to be developed. This is because unlike finite difference meshes, mesh lines do not intersect at right-angles. This means that high-order interpolation methods such as Bi-Cubic Spline techniques, commonly used to derive field information from finite difference meshes cannot be directly applied to finite element meshes. Several high-order interpolation methods have been devised to overcome this problem, three of which will be described in this section.

The greater flexibility of the mesh node layout in the finite element method as compared with finite difference meshes produces a wide class of different mesh types and high-order interpolation is critically dependent on mesh topology. A certain class of interpolation techniques apply to structured meshes only, while others have been developed for use on scattered mesh node data. The region block, graded, conformal, quasi-curvilinear, and boundary-fitted coordinate meshes are all of the structured type, while Delaunay triangles meshes are usually unstructured.

4.1 C^1 Triangle Interpolant

The first high-order interpolation technique considered is one reported by Renka and Cline [24], and can be found in the NAGLIB routines E01SAF/E01SBF [25]. This type of interpolation method is suitable for scattered node data, that is, for unstructured meshes. The method is based upon a triangle cubic interpolant which takes the function values and its derivatives at the mesh nodes as input data. It operates by dividing the triangle into three subtriangles with equal areas and uses a cubic function approximation in each subtriangle. The normal derivative of the function varies linearly along the outer edge. The function values and its partial derivatives along the edge depend only on the data values at the end points of the edge. The method yields C^1 continuity, that is it gives continuity of the function and its first derivatives over a mesh composed of triangle elements. The routine E01SAF first creates a triangle mesh from scattered data points by using Thiessian triangulation [26].

An important feature of the C^1 triangle interpolating scheme is that the partial derivatives of the function at the triangle vertices must first be specified. The routine E01SAF estimates these derivative values by using a local quadratic polynomial expansion fit that uses least squares weighting. Only the values of the potential at mesh nodes within a certain distance around the desired point are involved in the polynomial fit.

4.2   Normalised Hermite-Cubic interpolation

Another method of interpolation on finite element meshes is reported by Khursheed and is based upon using Hermite-Cubic interpolation on quadrilateral shaped elements [27, 23]. This method is only effective on structured meshes. It cannot be used on meshes whose lines are locally discontinuous. The concept of this approach is to map parts of the mesh into normalised space where mesh lines form a regular grid structure, and then use high-order interpolation methods in normalised space to calculate function derivatives. Hermite-Cubic interpolation is used for this purpose. These derivatives are then transformed to the desired spatial derivatives in real space by using the Jacobian of the transformation. Spline functions are used in normalised space to provide function derivatives and cross-derivatives at the mesh nodes for the Hermite-Cubic interpolation. The transformation from real space coordinates (x,y) into normalised coordinates (u,v) can be done globally or separately to each region.

If a mesh cell is bounded by the interval (u_1,u_2) and (v_1,v_2) in the u-v plane, then Hermite-Cubic interpolation for the potential f(u,v) can be expressed in the following form

$$f(u,v) = \sum_i^4 \sum_j^4 c_{ij} \left(\frac{u - u_1}{u_2 - u_1} \right)^{i-1} \left(\frac{v - v_1}{v_2 - v_1} \right)^{j-1} \qquad (11.12)$$

where c_{ij} are coefficients that depend on the values of the function f, its first-order partial derivatives f_u, f_v, and cross-derivative f_{uv} at the vertices of the quadrilateral element.

In the x-y plane, the function's spatial derivatives f_x and f_y are then found by

$$\begin{pmatrix} f_x \\ f_y \end{pmatrix} = J^{-1} \begin{pmatrix} f_u \\ f_v \end{pmatrix} \quad \text{where} \quad J = \begin{pmatrix} x_u & y_u \\ x_u & y_v \end{pmatrix} \qquad (11.13)$$

The derivatives x_u, x_v, y_u and y_v at a point (u,v) can be easily found analytically in the case of a distorted quadrilateral region block mesh. For an approximately smooth mesh, they can be found from Spline/Hermite-Cubic interpolation in the normalised plane: splines are first used to calculate the derivatives and cross-derivatives of x and y at all mesh nodes (with respect to the normalised coordinates). High-order interpolation in this way is used to construct smooth mesh lines between the mesh nodes.

Obviously an inverse mapping problem must first be made before derivatives of the function can be calculated in normalised space, since the Jacobian and function derivatives are dependent on u and v: that is for a given point (x_p, y_p), its corresponding position in the normalised plane, (u_p, v_p) must first be found. Let the mapping be denoted by functions $g_1(u,v)$ and $g_2(u,v)$ so that

$$x_p = g_1(u_p, v_p)$$
$$y_p = g_2(u_p, v_p)$$
(11.14)

Since these mapping functions are either known analytically or obtained numerically, the Newton Raphson iteration method can be used to find (u_p, v_p). To locate the mesh cell which contains the point (x_p, y_p), the Newton Raphson method is used in conjunction with the shape function for each element in the domain. If the shape functions are normalised so that u and v lie between 0 and 1 for each element, then the element which contains the point of interest will have non-negative (u,v) values which are less than unity, and can thus be easily identified. For the block region mesh, region blocks are first tested in this way to identify the region enclosing (x_p, y_p).

Another method to locate the mesh cell containing the point of interest is to use triangle element shape functions. These functions are explicitly expressed in terms of the (x,y) coordinates of the triangle vertices. Each triangle in the domain is tested in turn until the one that has all shape functions less than unity and greater than 0 is selected.

For the quadrilateral block region mesh, each mesh cell has a quadrilateral shape so that the mapping functions $g_1(u,v)$ and $g_2(u,v)$ are linear and of the following form

$$g_1(u, v) = a_0 + a_1 u + a_2 v + a_3 uv$$
$$g_2(u, v) = b_0 + b_1 u + b_2 v + b_3 uv$$
(11.15)

where coefficients a_0, a_1, a_2, a_3, b_0, b_1, b_2 and b_3 depend on the (x,y) coordinates of the mesh cell vertices.

For an approximately smooth mesh, Hermite-Cubic interpolation is used to construct smoothly varying mesh lines where the mapping functions are given by

$$g_1(u, v) = \sum_i^4 \sum_j^4 a_{ij} \left(\frac{u - u_1}{u_2 - u_1} \right)^{i-1} \left(\frac{v - v_1}{v_2 - v_1} \right)^{j-1}$$
(11.16)

$$g_2(u, v) = \sum_i^4 \sum_j^4 b_{ij} \left(\frac{u - u_1}{u_2 - u_1} \right)^{i-1} \left(\frac{v - v_1}{v_2 - v_1} \right)^{j-1}$$

here the coefficients a_{ij} depend on x, its derivatives (x_u, x_v) and cross-derivatives (x_{uv}) at the mesh cell vertices, and similarly the b_{ij} coefficients depend on y, its derivatives (y_u, y_v) and cross-derivatives (y_{uv}).

Since for the region block mesh the spatial derivatives are discontinuous at the edges of a region, an overlapping region scheme can be used to reduce the effect of the discontinuity. Figure 11.17 shows overlapping region lines. The potential values along these extended region lines are found from linear interpolation. Spline interpolation is then carried out within the u-v plane inside each extended region, providing greater continuity of the function's derivatives on the boundaries of the original region.

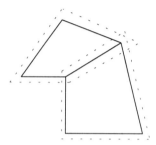

Figure 11.17 Overlapping regions

4.3   Laplace polynomial interpolation

Another high-order method of interpolation is a local polynomial fit to the potential at the mesh nodes which uses basis functions that satisfy Laplace's equation. This method has been presented by Chmelik and Bath for potential distributions which are solved in an axisymmetric cylindrical coordinate system [28]. The following implementation of their technique formulates it also for Laplaces's equation in two dimensional and three dimensional Cartesian coordinates. A general polynomial $P(x,y)$ expansion representing the local potential distribution is first considered, where powers up to x^n and y^n are included

$$P(x, y) = (a_0 + a_1 x + a_2 x^2 + \ldots a_n x^n)(b_0 + b_1 y + b_2 y^2 + \ldots b_n y^n) \quad (11.17)$$

$a_0 \ldots a_n$ and $b_0 \ldots b_n$ are coefficients of the expansion. The coordinates x and y are taken from the centre of the mesh cell which contains the point of interest and are normalised to the dimensions of the mesh cell.

The above polynomial is expanded out term by term, differentiated twice with respect to x and y, and substituted into Laplace's equation. This procedure will result in equations that link together the coefficients $a_0 \ldots a_n$ and $b_0 \ldots b_n$. By accounting for the relationship between these coefficients, $P(x,y)$ can then be simplified and expressed by a linear sum of basis functions all of which satisfy Laplace's equation. Let the basis functions be denoted by $g_i(x,y)$ and let c_i represent their coefficients, where i varies from 1 to N. $P(x,y)$ is expressed by

$$P(x, y) = \sum_{i=1}^{N} c_i g_i(x, y) \tag{11.18}$$

It is a straightforward matter to show that the first eleven terms of this polynomial are given by

$$g_1 = 1 \quad g_2 = x \quad g_3 = y \quad g_4 = xy \quad g_5 = x^2 - y^2 \quad g_6 = x^3 - 3y^2 x \quad g_7 = 3x^2 y - y^3$$
$$g_8 = x^3 y - xy^3 \quad g_9 = x^4 - 6x^2 y^2 + y^4 \quad g_{10} = x^5 - 10x^3 y^2 + 5xy^4$$
$$g_{11} = 5x^4 y - 10x^2 y^3 + y^5$$

$$\tag{11.19}$$

The basis functions used for the axisymmetric cylindrical coordinate system derived here are identical to the ones presented by Chmelik and Bath. They were found by substituting a general polynomial expansion for $P(r,z)$ into Laplace's equation in cylindrical coordinates. Here a point (r,z) is expressed in terms of the local coordinates (x,y) so that, $y=(r-r_A)/d$ and $x=(z-z_A)/d$, where (r_A, z_A) is the coordinate of the centre of the mesh cell in which the point (r,z) is located and d is equal to half the maximum mesh diagonal length. The first seven basis functions are given by

$$g_1 = 1 \quad g_2 = x \quad g_3 = -2yr_A + \left(2x^2 - y^2\right) \quad g_4 = -6r_A xy + x\left(2x^2 - 3y^2\right)$$
$$g_5 = -12r_A^2(x^2 - y^2) + 12r_A y\left(y^2 - 4x^2\right) + \left(8x^4 - 24x^2 y^2 + 3y^4\right)$$
$$g_6 = -20r_A^2(x^3 - 3y^2 x) + 20xy\left(3y^2 - 4x^2\right) + x\left(8x^4 - 40x^2 y^2 + 15y^4\right)$$
$$g_7 = 40r_A^3(3x^2 y - y^3) + 20r_A^2\left(-4x^4 + 21x^2 y^2 - 3y^4\right) + 30yr_A\left(-8x^4 + 12x^2 y^2 - y^4\right)$$
$$+ \left(16x^6 - 120x^4 y^2 + 90x^2 y^4 - 5y^6\right)$$

$$\tag{11.20}$$

Note that the ratio x/r_A or y/r_A becomes rapidly small as the point (r,z) moves away from the axis. For only a few mesh lines away from the axis, the terms involving powers of r_A will become comparatively large, and the seven basis functions for the axisymmetric cylindrical coordinates will tend towards becoming the first seven basis functions given for rectilinear coordinates (x,y). For reasons of better numerical stability, it was found necessary to divide each basis function by its leading term.

The coefficients c_i are solved by fitting $P(x,y)$ to the potential at a number of mesh nodes around the point of interest. If M mesh nodes are considered, then obviously M must be greater than equal to N for the resulting M system of equations to have a solution. The system of equations are put in the following form

$$
\begin{pmatrix} w_1 P_1 \\ w_2 P_2 \\ . \\ . \\ . \\ w_M P_M \end{pmatrix} = \begin{pmatrix} w_1 g_1(x_1,y_1) & w_1 g_2(x_1,y_1) & . & . & w_1 g_N(x_1,y_1) \\ .w_2 g_1(x_2,y_2) & .w_2 g_2(x_2,y_2) & . & . & .w_2 g_N(x_2,y_2) \\ . & & . & . . & . \\ . & & . & . . & . \\ . & & . & . . & . \\ w_M g_1(x_M,y_M) & w_M g_2(x_M,y_M) & . & . & w_M g_N(x_M,y_M) \end{pmatrix} \begin{pmatrix} c_1 \\ c_2 \\ . \\ . \\ . \\ c_N \end{pmatrix}
$$

$$(11.21)$$

where w_1 .....w_M are weighting factors which are chosen to give more importance to the nodes closer to the point (x_p, y_p).

To solve for the coefficients c_1......c_N, a singular value decomposition technique was used [29]. This method can monitor the condition number of the system matrix and allows for the suppression of rows that lead to ill-conditioning. Singular value decomposition involves expressing the inverse of a matrix as $[V]\,[D]\,[U]^T$ where $[D]$ is a diagonal matrix. The condition number is defined to be the ratio, $\min(d_j)/\max(d_j)$ for $j=1$ to N, where N is the number of entries in $[D]$. In general, the numerical noise present at the mesh nodes should be smaller than the condition number for stable results.

To locate the mesh cell that contains the point (x_p, y_p), Newton-Raphson iteration was used in conjunction with each element's shape function.

In three dimensions, a polynomial $P(x,y,z)$ is expanded in terms of N basis functions $g_i(x,y,z)$, $i=1$ to N, which all individually satisfy Laplace's equation. Consider the following polynomial function expansion

$$f(x,y,z) = a_0 + a_1 x + a_2 y + a_3 z + a_4 xy + a_5 xz + a_6 yz + a_7 xyz + ...a_{63} x^3 y^3 z^3$$

$$(11.22)$$

This polynomial is generated by multiplying together three third order polynomials, each one representing a different direction. The basis functions $g_i(x,y,z)$ can be found from substituting this polynomial into Laplace's equation

$$f_{xx} + f_{yy} + f_{zz} = 0 \qquad\qquad (11.23)$$

This relates the unknown 64 coefficients $a_0....a_{63}$ together. It is straightforward to show that the 64 unknown coefficients will then reduce to $c_0....c_{25}$ coefficients, each multiplying a corresponding basis function.

These functions are

$g_1 = 1 \quad g_2 = x \quad g_3 = y \quad g_4 = z$ (1st-order)

$g_5 = x^2 - y^2 \qquad g_6 = x^2 + y^2 - 2z^2 \qquad g_7 = xy \qquad g_8 = yz$

$g_9 = xz$ (2nd-order)

$$g_{10} = xyz \quad g_{11} = x^2y - (1/3)y^3 \qquad\qquad g_{12} = x^2z - (1/3)z^3$$
$$g_{13} = y^2x - (1/3)x^3 \qquad\qquad\qquad g_{14} = y^2z - (1/3)z^3$$
$$g_{15} = z^2x - (1/3)x^3 \qquad\qquad\qquad g_{16} = z^2y - (1/3)y^3 \qquad (3^{rd}\text{-order})$$

$$g_{17} = x^4 - 3x^2y^2 - 3x^2z^2 + 3y^2z^2 \qquad g_{18} = y^4 - 3x^2y^2 + 3x^2z^2 - 3y^2z^2$$
$$g_{19} = z^4 - 3x^2y^2 - 3x^2z^2 - 3y^2z^2 \qquad g_{20} = x^3y - 3z^2xy$$
$$g_{21} = x^3z - 3y^2xz \qquad\qquad\qquad\quad g_{22} = y^3x - 3z^2xy$$
$$g_{23} = y^3z - 3x^2yz \qquad\qquad\qquad\quad g_{24} = z^3x - 3y^2zx$$
$$g_{25} = z^3y - 3x^2zy \qquad\qquad\qquad\qquad\qquad\qquad\qquad (4^{th}\text{-order})$$

$$(11.24)$$

A total of 32 potential values were used in the Laplace polynomial fit. The first eight points come from the corner nodes of the mesh cell which encloses the point of interest, while a further 24 points come from nodes on neighbouring mesh cells, as shown in Figure 11.18. A weighted least square fit to the 32 points is made where a weighting factor can be selected to vary the importance given to each point. In general, greater importance is given to points closer to the field evaluation point, while those further away should be assigned a smaller weighting factor. The first eight points are assigned a weighting factor of unity, while the other 24 points are assigned a weighting factor of 0.05. Weighting factors much higher than 0.05, say over 0.2, caused instabilities for points close to a conductor surface, and for this reason, they were not used.

It was also found that only the first 16 basis functions were required in the polynomial fit. Not only did use of all 25 functions cause significant errors close to conductor surfaces, but it also made the condition number much smaller, typically by an order of magnitude, and the results were in general worse than for a 16 function polynomial fit.

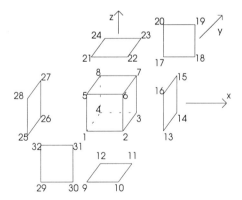

Figure 11.18 The local node scheme used for Laplace polynomial interpolation in three dimensions.

5.   **FLUX LINE REFINEMENT FOR THREE DIMENSIONAL
        ELECTROSTATIC PROBLEMS**

In general, the technique of modelling curved boundaries by creating local distortions on a rectilinear mesh will produce a relatively low number of nodes that lie on electrode surfaces. For this reason, both the truncation error generated within the finite element program and the error incurred through high-order field interpolation is expected to be high close to electrode surfaces. Obviously a mesh refinement technique is required for evaluating the electric field near electrodes.

One method is to generate a structured refinement mesh from the equipotential and flux lines of the initial potential solution. In three dimensions however, due to the difficulty of automatically generating a suitable coordinate system for each equipotential plane, it is easier to refine the potential along flux lines. Mesh refinement along a flux line involves solving a one dimensional field problem and is thus a relatively inexpensive computational task. Flux lines can be plot from the background potential distribution in a variety of different ways. The most straightforward method is to plot trajectories that are perpendicular to equipotential surfaces and which start from an electrode surface. Electric field values can then be found by using Cubic-Spline interpolation along the flux line. A locally structured mesh near an electrode surface can be created from several flux lines, and the electric field from points on this mesh can be found from Bi-Cubic interpolation.

Consider an orthogonal three dimensional coordinate set (u,v,w) where the flux line is propagating in the u direction. Since at every point on the flux line v and w are tangential to an equipotential plane, then the derivatives of the electric potential $\Phi(u,v,w)$ with respect to w and v will be zero. It follows that the solution to Laplace's equation reduces to solving the following one dimensional equation,

$$\frac{\partial}{\partial u}\left(\frac{h_2 h_3}{h_1}\Phi_u\right)=0 \quad \text{where} \quad \begin{aligned} h_1 &= \left[x_u^{\,2}+y_u^{\,2}+z_u^{\,2}\right]^{1/2} \\ h_2 &= \left[x_v^{\,2}+y_v^{\,2}+z_v^{\,2}\right]^{1/2} \\ h_3 &= \left[x_w^{\,2}+y_w^{\,2}+z_w^{\,2}\right]^{1/2} \end{aligned} \qquad (11.25)$$

To calculate the derivatives of x, y and z with respect to u, v and w, flux lines can be plot which start at the positions Δw and Δv to either side of the flux line under consideration, as shown in Figure 11.19. Here, the central flux line labelled 0, is surrounded by flux lines 1 and 3 in the v direction and 3 and 4 in the w direction. The coordinates for points at the same potential along these flux lines can be readily found. In this way, the derivatives of x, y and z with respect to u, v and w along equipotential planes can be speedily calculated.

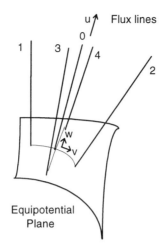

Figure 11.19 Flux line mesh refinement

6.   ACCURACY TESTS

6.1   High-order interpolation in Two Dimensions

To assess the performance of high-order interpolation methods, fortran programs for each method described in section 4 were written by the author, except in the case of the C^1 triangle interpolant where the E01SAF/E01SBF NAGLIB routines were used.

The (u_p, v_p) point in normalised space for mesh cell identification and for use with the Spline/Hermite-Cubic interpolation method was typically found to 6 digit precision in less than 8 Newton-Raphson iterations. For the test potential distributions considered, this level of imprecision on the (u_p, v_p) coordinates was found to be adequate.

Normally eleven functions with twelve mesh nodes were used for the Laplace polynomial expansion interpolation method in two-dimensional rectilinear coordinates. However, this led to large errors when the point of interest lay near a conductor boundary or where the truncation noise was high. For these situations, only seven functions were used. Dropping a row out of the function fit was done where the condition number fell to less than 10^{-6}. A weighting factor of unity was used for the four closest nodes to the point to be evaluated, while for other nodes, a weighting factor of 0.2 was used.

The closed box conductor layout shown in Figure 11.20a was used to provide a test case to measure interpolation accuracy on the region block mesh type. All the conductors are put to 0 volts except for the top plate which is specified to be 100 volts. The width of the box is chosen to be 10 cm, one half of its height. The analytical solution for the potential distribution of this electrostatic structure can easily be derived from the Fourier series method [30].

The four quadrilateral regions depicted in Figure 11.20b were evenly subdivided to generate the 21 by 41 mesh shown in Figure 11.20d. The point P denotes an internal region corner node, and lies at the coordinate x=5cm and y=7cm. Its position is obviously related to the degree of distortion on all regions. Interpolation is performed to provide electric field values at points lying on lines A and B shown in Figure 11.20c.

To investigate the errors related to the field interpolation procedure itself, analytical potential values at the mesh nodes were used. In all cases considered, an eleven basis function fit in the Laplace polynomial expansion gave more accurate results than a 7 basis function fit.

The results are first carried out on the 21 by 41 mesh lines shown in Figure 11.20d.

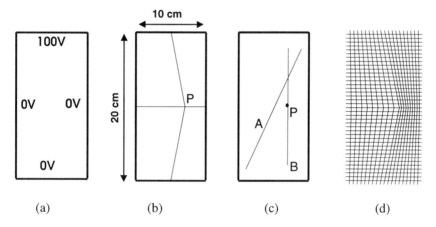

Figure 11.20 2D Electric Box test example for interpolation in (x,y) plane

Figure 11.21 shows the accuracy of the electric field strength for points on line A using the Hermite-Cubic method and C^1 triangle interpolant. The results show that the Hermite-Cubic method is typically two orders of magnitude more accurate than the C^1 triangle interpolant. The average error on the electric field strength for the Hermite-Cubic method is approximately 0.006%.

Figure 11.22 compares together the performance of the Hermite-Cubic and the Laplace polynomial expansion methods for exactly the same conditions. The average error on the electric field strength for the Laplace polynomial expansion method is around 0.0007%, an order of magnitude lower than that for the Hermite-Cubic method at the same points. This shows that of the three types of high-order interpolation methods considered, the Laplace polynomial expansion method of interpolation is the most accurate.

However, for the next two points along line A, not shown on the graph for reasons of scale, the error of the Laplace polynomial expansion method rose to over a value of 0.23%, a factor of over two greater than Hermite-Cubic interpolation for the same points. The sharp rise in error for these two points comes from rapid changes in the potential function value. This result shows that the errors of the Laplace polynomial expansion method are more sensitive to large field changes than the Hermite-Cubic method of interpolation. This inaccuracy at points of rapid field change is not a problem of matrix ill-conditioning, since for the same point using a seven basis function fit, the error rose to over 0.42% with a condition number 20 times lower than for the eleven function fit.

When a first-order finite element solution was used, the three high-order methods gave approximately the same value, which varied between 0.1 to 1%, as shown in Figure 11.23. This shows that the final accuracy is dominated by truncation error. However, the results are still an order of magnitude better than linear interpolation for the same conditions.

The fact that a significant improvement is not obtained for the C^1 triangle interpolant when analytical potential values at the mesh nodes are used shows that this method is not a very accurate technique compared with the other methods of high-order interpolation. The weakness of the method lies in estimating the function's derivatives at the vertices of each triangle. This cannot be done accurately on a triangle mesh and the procedure of using a local quadratic polynomial fit limits the accuracy of the overall method.

To assess the effect of geometrical distortion of the mesh on interpolation accuracy, electric field values were calculated around point P on line B (see Figure 11.20c). Analytical values for the potential were used at the mesh nodes. The results are depicted in Figure 11.24 for the Hermite-Cubic and Laplace polynomial expansion methods of interpolation. They show that for the Hermite-Cubic method, the error on the field value rises sharply at point P, and is around an order of magnitude worse than for the other points along the line. The Laplace polynomial expansion method however, does not seem to be significantly degraded by the mesh distortion. The average error of Hermite-Cubic interpolation along the line was around 0.004%, while the Laplace polynomial expansion method gave an average error of around 0.00067%.

This result demonstrates an important advantage of the Laplace polynomial expansion method. Where the truncation error is low, it performs well at region corners and boundaries of the region block mesh. Of course, distortion will generally produce greater truncation noise at the mesh nodes, which will contribute to greater imprecision on final field values.

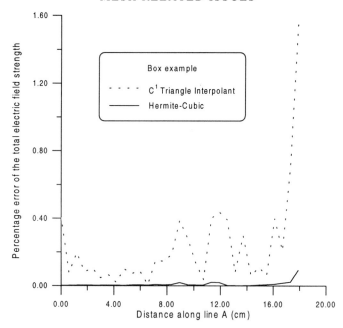

Figure 11.21 Comparison of the Hermite-Cubic and C^1 Triangle Interpolant along line A on the box example for analytical potential values at the mesh nodes

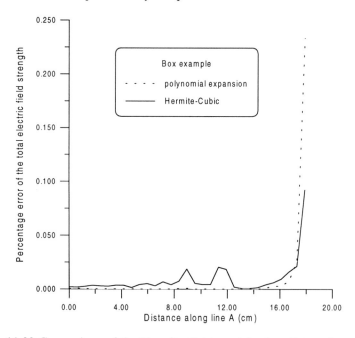

Figure 11.22 Comparison of the Hermite-Cubic and Laplace Interpolant along line A on the box example for analytical potential values at the mesh nodes

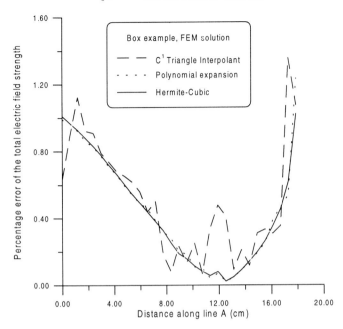

Figure 11.23 Comparison of different high-order interpolation methods along line A on the box example for FEM solved potentials

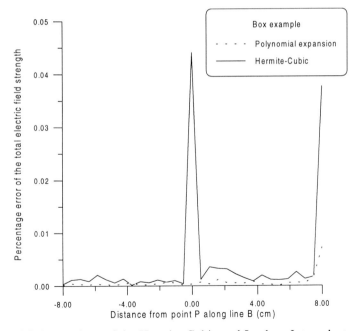

Figure 11.24 Comparison of the Hermite-Cubic and Laplace Interpolant along line B on the box example for analytical potential values at the mesh nodes

The sensitivity of the Hermite-Cubic method to mesh distortion at point P obviously comes from the discontinuity of the fields at the region block edges. Although the overlapping region technique reduces this continuity, further improvement is required. One step might be to use a larger overlapping area. In the present interpolation routine the regions overlap only by one mesh cell.

Further investigations on finer meshes reveals that the Hermite-Cubic interpolation error due to mesh distortion decreases approximately by an order of magnitude as the mesh resolution is doubled. For a mesh resolution of 41 by 81, the peak value of the error decreased to around 0.005%, as shown in Figure 11.25. For a mesh resolution of 81 by 161, the inaccuracy at point P was almost entirely eliminated, and the average error fell to below 0.001%.

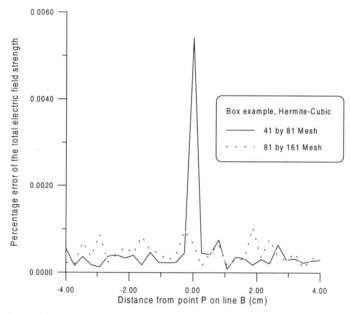

Figure 11.25 Comparison of Hermite-Cubic interpolation along line B on the box example for analytical potential values at the mesh nodes at different mesh resolutions

One important observation of the Laplace polynomial technique is that its performance is critically dependent on the angular distribution of the mesh nodes surrounding the point of interest. If for instance, the aspect ratio of the mesh cells is large, the condition number for the polynomial fit rises significantly and produces considerable loss of accuracy. Take for instance the situation where the box height is changed from 20 cm to 2 cm, keeping the box width fixed at 10 cm. Here the aspect ratio of the mesh cells is changed from approximately unity to 10. For an 11 function fit, the condition number goes from around 1/17 to 1/82,000, which results in more than an order of magnitude loss of accuracy. The results are now worse than those produced by the Hermite-Cubic interpolation method.

There are at least two possible ways to remedy the situation. One approach might be to use less basis functions. When a 7 function fit was made, the condition number went from 1/82,000 to 1/60. However, the final result, although more stable, is obviously going to be less accurate since the basis functions used in the fit are of lower order. The accuracy in this case dropped by an order of magnitude, and was worse than for Hermite-Cubic interpolation. Another possibility is to use mesh nodes that are more evenly distributed (assuming that they are available).

In principal, higher-order elements can always be used to reduce the truncation noise to levels that are comparable to interpolation noise. For a 41 by 81 mesh on the box example shown in Figure 11.20, the accuracy of a second-order element solution was only marginally worse than the one obtained by using the analytical potentials at the mesh nodes. These results were produced by the Laplace polynomial expansion technique and are shown in Figure 11.26. The errors on the total field strength range from 0.5×10^{-3} to $4 \times 10^{-3}\%$, and are more than adequate for most applications in charged particle optics.

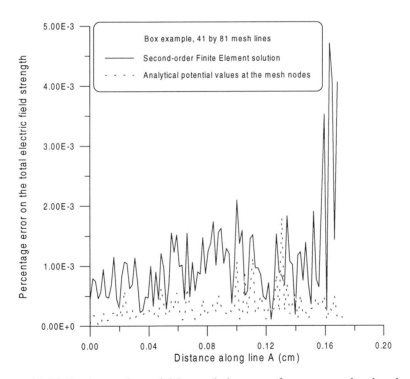

Figure 11.26 Laplace polynomial interpolation error from a second-order element solution along line A for the box example

It should be noted that in many cases, Richardson's extrapolation technique can also be used to reduce truncation noise levels, however, it is only effective if mesh lines intersect each other at right-angles and if the interpolation point is not close electrodes.

The accuracy of higher-order interpolation in axisymmetric (r,z) coordinates was investigated for the two-tube lens example shown in Figure 8.2. Interpolation accuracy was investigated by using analytically derived potentials at the mesh nodes. Here, as for the x-y plane, the Hermite-Cubic and Laplace interpolation methods were found to be several orders of magnitude more accurate than Linear interpolation and the C^1 triangle interpolant. Although the Laplace polynomial technique in (r,z) is the most accurate of the methods examined, it exhibits signs of being unstable. For mesh cells either having large aspect ratios or which are close to electrode boundaries, the condition number can easily be lower than 10^{-7}. For the region block mesh shown in Figure 8.2b, where at r=0, z=0 the mesh cell aspect ratio is 10, the condition number falls to 3.3×10^{-7}, producing errors of greater than 10,000%. On the other hand, for r=0.1 and z=0.1, where the aspect ratio is 1.25, an error of 1.5×10^{-6} % on the total electric field strength is obtained. The corresponding error for Hermite-Cubic interpolation is 2.1×10^{-5}%.

The Hermite-Cubic inaccuracy at the region block edges was not found to be significant (for the region block mesh shown in Figure 8.2b). The level of errors on the total electric field strength around the centre of the lens typically ranged from 10^{-6} to 10^{-4}%. The effect of such errors on the primary beam electron trajectory path is small. This can be demonstrated by investigating their effect on the calculation of the 3rd and 5th spherical aberration coefficients derived by direct ray tracing. Using 51 parallel trajectories entering the lens at input radii evenly spaced between 0 and 0.6R (R is the lens tube radius), then the values for C_{s3} and C_{s5} are calculated to be 9.98701 and 106.523 respectively. Here analytical potentials are used at the mesh nodes and region block Hermite-Cubic interpolation is used to derive the electric field values. Details of how the aberration coefficients are calculated by direct ray tracing for this simple test lens example are discussed in section 2 of Chapter 8. The above values can be compared to the situation where no mesh is used, that is, where electric fields are used directly from the analytical formula. In this latter case, C_{s3} is found to be 9.9872 and C_{s5} is 106.587. The errors related to region block Hermite-Cubic interpolation are then estimated to be around 1.9×10^{-3}% for C_{s3} and 0.06% for C_{s5}. These errors are relatively small, and are typically one to two orders of magnitude smaller than those obtained from first-order finite element solutions.

It is also important to establish the accuracy of high-order field interpolation for electron gun simulation. Here, unlike the situation for electron lenses, electron trajectory paths must be traced very close to conductor boundaries. A suitable test example for this purpose is the parabolic capacitor layout, shown in Figure 11.27a. Mesh lines are defined by the well-known parabolic coordinates (u,v) and are shown in Figure 11.27b. The quasi-conformal electron gun mesh shown in Figure 11.15 is topologically similar to this test example. Here, there are 24 lines in the u-direction and 41 mesh lines in the v-direction. The coordinate v varies from 1 to 10, where the cathode at v=1.0 is set to a potential of 0 volts, and the anode located at v=10 is set to a potential of 1 volt. Both electrodes extend from the axis where the coordinate u=0, to the right hand side boundary where u=5.0. The normal component of the potential gradient at the right hand side boundary is zero (Neumann boundary).

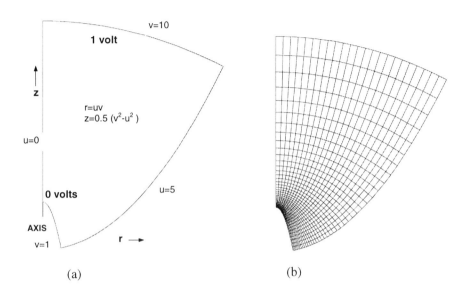

(a)   (b)

Figure 11.27 Parabolic Capacitor test example
(a) Electrode layout
(b) 24 by 41 mesh lines

The mesh lines shown in Figure 11.27b were generated according to the equipotential lines and flux lines of the parabolic capacitor field solution in axisymmetric cylindrical coordinates (r,z). The potential in the analytical solution for the parabolic capacitor has the well known logarithmic variation with the coordinate v. To assess the accuracy of the global Hermite-Cubic interpolation method on this mesh, analytical potential values were specified at the mesh nodes and the normal electric field at the centre of the cathode tip (u=0, v=1) was calculated. A value of 0.434162 V/m was obtained, which has an error of around 0.03% when compared to the analytical value of 0.43429448 V/m. When the mesh resolution is doubled (to 47 by 81 mesh lines), this error falls by an order of magnitude, to around 0.0038%. These kinds of error levels are relatively small and do not significantly affect the electron trajectory paths electrons emitted from the cathode.

Field interpolation under the same conditions from a first-order finite element solution is significantly worse. This is because the accuracy of the calculated field value in this case is limited not by interpolation noise, but by truncation noise. For the 24 by 41 mesh, an error of around 0.164% is obtained, while for the 47 by 81 mesh, the error was found to be 0.0328%. Although Richardson's extrapolation further lowered this error to around 0.025%, it is still obviously limited by truncation noise. However, second-order finite element solutions on the same mesh approaches the interpolation error limit. For the 24 by 41 mesh, an error of 0.0475% is obtained from the second-order finite element solution. This is close to the error obtained for analytical potentials at the mesh nodes (0.03%). On the 47 by 81 mesh, the second-order finite element solution provides an error of 0.006%, which is only a factor of 1.6 higher than the interpolation limit (0.0038%).

6.2   The Ampere's Circuital mesh test

One useful way of monitoring the field accuracy of finite element meshes is to use Ampere's Circuital law for each mesh node. In the lens bore region (current free), all closed paths of $\mathbf{H} \cdot \mathbf{dl}$ should be zero. Lencová has proposed this to provide a global estimate of the axial field error [31], where $\mathbf{H} \cdot \mathbf{dl}$ is evaluated around the outer boundary of the solution domain. A similar method can also be used at the local mesh level, where $\mathbf{H} \cdot \mathbf{dl}$ contributions are calculated for each side of an element. Around a mesh cell whose sides 1, 2, 3 and 4 are bounded by the vectors $\mathbf{dl}_1$, $\mathbf{dl}_2$, $\mathbf{dl}_3$ and $\mathbf{dl}_4$ respectively, the mesh current error ΔI is given by

$$\Delta I = \int_1 \mathbf{H}_1 \cdot \mathbf{dl}_1 + \int_2 \mathbf{H}_2 \cdot \mathbf{dl}_2 + \int_3 \mathbf{H}_3 \cdot \mathbf{dl}_3 + \int_4 \mathbf{H}_4 \cdot \mathbf{dl}_4 \qquad (11.26)$$

where the vectors $\mathbf{H}_1$, $\mathbf{H}_2$, $\mathbf{H}_3$ and $\mathbf{H}_4$ represent magnetic field intensities along the sides of the mesh cell (element). The finite element method provides values of the vector potential A at each mesh node, and high-order interpolation is used to derive the components of $\mathbf{H}$ along the sides of each mesh cell.

The single-pole lens shown in Figure 11.1 is used as a test example, and the mesh current error distribution ΔI in the lens bore is calculated for three different mesh types: a region block mesh, a graded mesh, and a quasi-conformal mesh. The near pole-piece region for each of these mesh layouts is shown in Figures 11.28 a-c. Each of these mesh layouts has around 20 lines from the optic axis to the single pole-piece. To make a fair comparison, all mesh current errors are carried out on the conformal mesh, where $\mathbf{H}$ values for each mesh solution are derived from high-order field interpolation and the $\mathbf{dl}_1 \ldots \mathbf{dl}_4$ vectors are defined on conformal mesh cells. The final mesh current error is generated both by interpolation and mesh truncation noise. The lens bore region for the mesh current error test is shown in Figure 11.28c. It should be noted that a divergence of $\mathbf{B}$ test can also be done, and the results are expected to be similar to the mesh current error test. The divergence of $\mathbf{B}$ test however is more general, since it can be applied throughout the whole domain. The $\mathbf{H.dl}$ test in comparison needs to be carefully formulated for mesh cells whose sides lie on air/iron interfaces. The mesh current error distribution ΔI derived from first-order and second-order solutions on the trial mesh are shown in Figures 11.29a and

11.29b, while Figures 11.30a and 11.30b present similar results for the graded and quasi-conformal meshes.

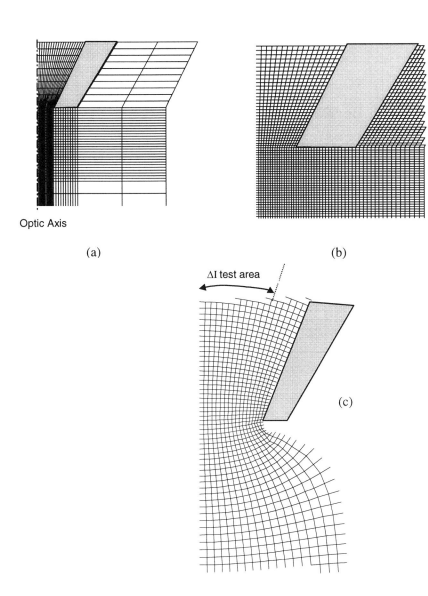

Figure 11.28 Near pole-piece mesh regions of a single-pole objective lens
(a) Region block
(b) Graded
(c) Quasi-conformal

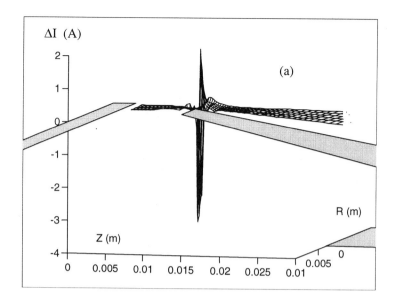

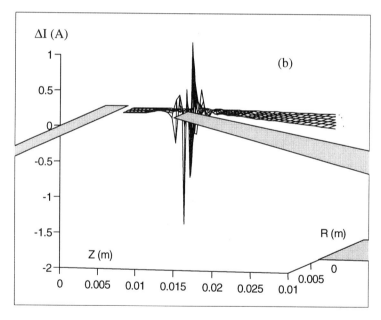

Figure 11.29 Mesh current error distribution derived from single-pole lens trial
mesh solutions
(a) First-order elements
(b) Second-order elements

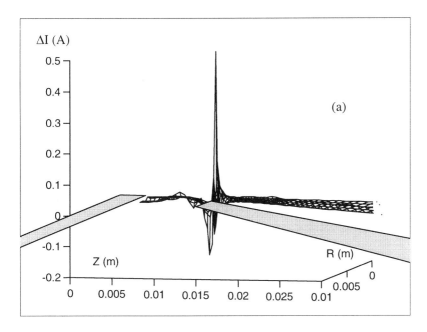

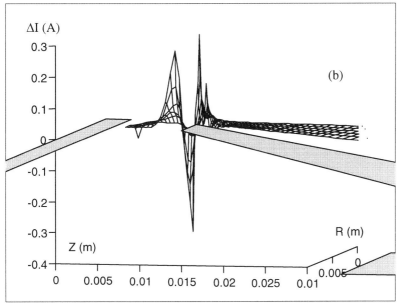

Figure 11.30 Mesh current error distribution derived from single-pole lens first-order refinement solutions
(a) Graded Mesh
(b) Quasi-conformal mesh

Hermite-Cubic region block interpolation is used on the trial mesh, while for the refinement meshes, a global Hermite-Cubic method is used. Figures 11.29a and 11.29b show that the peak mesh current error of the second-order element solution on the trial mesh is around half that of the corresponding first-order solution. Although the error is substantially reduced around the axis region, the far off-axis values around the pole-piece are dominated by interpolation errors, so the overall improvement is not significantly better. The error for the trial mesh is dominated by sharp changes in mesh line direction and spacing close to the tip of the pole-piece. On the other hand, mesh refinement is an effective way of reducing the interpolation error. The peak error for the graded mesh is a factor of 5 smaller than the trial mesh solution, while for the quasi-conformal mesh, it is around an order of magnitude smaller. For a given number of mesh nodes in the refinement mesh, the quasi-conformal mesh provides the lowest error. Based upon these results, it is recommended that quasi-conformal mesh refinement be used in combination with global Hermite-Cubic interpolation to plot the trajectories of secondary and backscattered electrons which travel up the objective lens bore.

Figures 11.31a and 11.31b show the trajectory paths of low energy secondary electrons (3 eV) and backscattered electrons (9 keV) which start at a distance of 2 mm below the single-pole piece tip. These electrons are emitted over a wide range of initial angles.

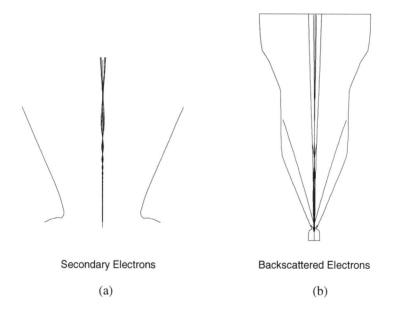

Secondary Electrons   Backscattered Electrons

(a)   (b)

Figure 11.31 Through-the-lens trajectories for the single-pole objective lens
(a) Secondary electrons with an energy of 3 eV
(b) Backscattered electrons with an energy of 9 keV

6.3   High-order interpolation in three dimensions

In three dimensions, field interpolation was carried out for the spherical capacitor test example shown in Figure 11.12. The results of interpolating for the electric field strength are shown in Figure 11.32. Laplace polynomial interpolation is compared to linear interpolation (using an element's first-order shape functions). The percentage error of the electric field strength is plotted as a function of radius for the spherical coordinate angles of $\theta=\pi/4$ and $\phi=1.2$ radians. The radial interval for which the electric field was computed lay between 2.1cm and 7.9cm. As expected, the percentage error is very high close to the electrode surface. At the first radial point, only 0.1cm away from the inner conductor surface, the percentage error is around 33%. This error is of course, even larger for points closer to the conductor boundary. On the other hand, in the interior region, where mesh cells are approximately rectangular, the error on the electric field strength does not exceed 3%. This level of accuracy is around an order of magnitude better than evaluating the electric field strength by linear interpolation.

The results were repeated for radial lines in a wide range of different directions and the inaccuracy close to the conductor on the initial mesh varied significantly. For $\theta=\pi/2$ and $\phi=0$ radians, the percentage error at the first radial point (r=2.1cm) dropped to 7% while for $\theta=\pi/4$ and $\phi=\pi/4$ radians, it dropped to less than 3%. Of course, in all cases, the error dramatically rose as the evaluated point approached the conductor surface.

Refinement along flux lines solves the problem of high inaccuracy close to the electrode boundary, as shown in Figure 11.33. The inaccuracies on the electric field strength typically fell from over 30% to less than 1%. The potential was resolved at 60 points along the flux line by finite difference in combination with successive over-relaxation. Only five equipotential surfaces were locally constructed, and the derivatives of x, y and z for the other points along the flux line were found by cubic spline interpolation. Not only were the results from refinement along the flux line consistently better in all directions, typically less than 1%, but the error did not become significantly large as the evaluation point approached the electrode surface. These results show that the method of mesh refinement along flux lines is an effective way of improving the accuracy of electric field computations close to curved electrode boundaries in three dimensions.

To summarise the results of this section: accurate field interpolation on finite element solved potential distributions can be performed in two and three dimensions. Structured mesh refinement greatly enhances the accuracy of field interpolation. Where this kind of mesh refinement is used, field errors from high-order element solutions approach the interpolation limit.

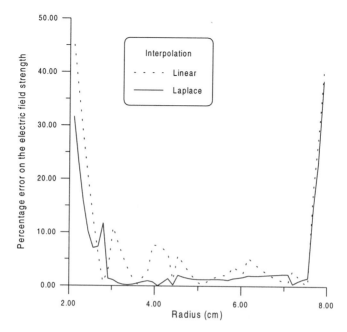

Figure 11.32 Accuracy of the electric field strength with radius for the spherical capacitor ($\theta=\pi/4$ and $\phi=1.2$ radians)

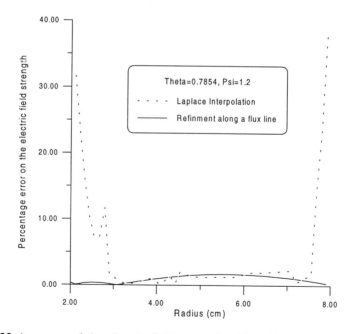

Figure 11.33 Accuracy of the electric field strength with radius for the spherical capacitor after flux line mesh refinement ($\theta=\pi/4$ and $\phi=1.2$ radians)

The following conclusions can be drawn from the accuracy tests presented in this book. On finite element solved potential distributions, there are three main sources of error which limit the precision of the direct ray tracing of charged particles: truncation errors, field interpolation errors and trajectory integration errors. Traditionally, mesh truncation and field interpolation errors dominate the overall accuracy to which trajectory paths can be plot. However, the results presented here show that this need not be the case. Both truncation and field interpolation errors can be made comparable to trajectory integration error levels. To reduce the truncation noise, high-order elements or Richardson's extrapolation method can be used, while high-order interpolation in combination with structured mesh refinement brings down the magnitude of field interpolation errors.

REFERENCES

[1]   "Numerical Grid Generation in Computational Fluid Dynamics and Related Fields", edited by N. P. Weatherill, P. R. Eiseman, J. Hauser, J. F. Thompson, Proc. of the 4th International Conference, Pineridge Press Ltd, Wales, UK, 1994

[2]   E. Munro, "Computer-aided design methods in electron optics", PHD Dissertation, University of Cambridge, Cambridge, UK, 1971

[3]   N. K. Kang, J. Orloff, L. W. Swanson and D. Tuggle, "An improved method for numerical analysis of point electron and ion source optics", J. Vacuum Science and Technology, 19(4), Nov./Dec. 1981, p1077-1086

[4]   B. Lencová, "Compuation of electrostatic lenses and multipoles by the first-order finite element method:, Nuclear Instruments and Methods in Physics Research A, 363, 1995, p190-7

[5]   K. J. Binns, P. J. Lawrenson and C. W. Trowbridge, "The analytical and numerical solution of electric and magnetic fields", John Wiley and sons, NY, 1992, section 13.3.4, p391-6

[6]   P. L. George, "Automatic mesh generation: application to finite elements", John Wiley and sons\Masson, Paris, 1991, chapter 11

[7]   A. Khursheed, "Mesh refinement for the finite element method in electron optics", Optik, 107, No. 3, 1998, p99-108

[8]   A. Khursheed, "The Boundary-fitted Coordinate Method to Improve the Numerical Analysis of Electron Optical Systems", Scanning, Vol. 16, 1994, p201-8

[9]   Z. Shao and P. S. D. Lin, "High resolution low voltage electron optical system for very large specimens, Rev. Sci. Instrum., 60 (11), p3434-3441, 1989

[10]   E. Munro, X. Zhu and J. Rouse, "High-order and multipole aberrations by aberration integral and direct ray-tracing methods", J. Micros., 179, p161-9, 1995

[11]   A. Khursheed, "The Khursheed Electron Optics Software" (KEOS), Department of Electrical Engineering, National University of Singapore, Singapore 119260, 1998

[12]   OPERA2D, Vector Fields Ltd, 24 Bankside, Kidlington, Oxford OX5 1JE, UK

[13]   J. F. Thompson, and Z. U. A. Warsi, "Boundary-fitted coordinate systems for numerical solution of partial differential equations – a review", J. Compuational Physics, 47, 1982, p1-108

[14] T. Takagi, K. Miki and H. Sano, "A new approach to electron beam analysis using a domain decomposition and overlapping method in a three dimensional boundary-fitted coordinate system", Appl. Num. Math 3, p305-316, 1987

[15] J. F. Thompson F. C. Thames, and C. W. Mastin, "Automatic numerical generation of body-fitted curvilinear coordinate system for field containing any number of arbitrary two-dimensional bodies", Journal of Computational Physics, 15, 1974, p299-319

[16] C. W. Mastin and J. F. Thompson, "Transformation of three-dimensional regions onto rectangular regions by elliptic equations", Numer. Math. 29, 1978, p397-407

[17] X. Zhu, and E. Munro, A computer program for electron gun design using second-order finite-elements, J. Vac. Sci. Technol. B7 (6), no. 11/12, p1862-1869, 1989

[18] A. Khursheed, "Curvilinear finite element mesh generation for electron gun simulation", Scanning, 19, 1997, p300-309

[19] P. W. Hawkes and E. Kasper, "Principles of Electron Optics", Vol. 1, Academic Press, 1989, chapter 13, p188-198

[20] J. L. Coulomb and G. Meunier, "Finite element implementation of virtual work principle for magnetic or electric force and torque computation", IEEE Transactions on Magnetics, Vol. 20, No. 5, p1894-1896, 1984

[21] S. Ratnajeevan H. Hoole and S. Subramaniam, "Higher finite element derivatives for the quick synthesis of electromagnetic devices", IEEE transactions on Magnetics, Vol. 28, No. 2, 1992

[22] P. W. Hawkes and E. Kasper, "Principles of Electron Optics", Vol. 1, Academic Press, 1989, chapter 13, p188

[23] A. Khursheed and A. R. Dinnis, "High accuracy trajectory plotting through finite element fields", J. Vac. Sci. Technol. B7(6), p1882-5, 1989

[24] R. L. Renka and A. K. Cline, "A Triangle-based C^1 interpolation method", Rocky Mountain J. Math., vol. 14, p223-7, 1984

[25] E01SAF/E01SBF NAG Mark 13, The Numerical Algorithms Group (NAG), Wilkinson House, Jordan Hill Road, Oxford OX2 8DR, England, United Kingdom.

[26] A. K. Cline and R. L. Renka, "A storage efficient method for construction of a Thessian triangulation, Rock Mountain J. Math., Vol. 14, p119-139, 1984

[27] A. Khursheed, "The Boundary-fitted coordinate method to improve the numerical analaysis of electron optical systems", Scanning, vol. 16, p201-208, 1994

[28] J. Chmelik and J. E. Barth, "An interpolation method for ray tracing in electrostatic fields calculated by the finite element method", Proc. SPIE, vol.2014 (Charged Particle Optics), editors W. B. Thompson, M. Sato and A. V. Crewe, p133-143

[29] W. H. Press, B. P. Flannery, S. A. Teukolsky and W. T. Vetterling, "Numerical Recipes", Cambridge University Press, Cambridge, 1986, p52-64

[30] E. Kreyszig, "Advanced engineering mathematics", John Wiley and Sons, 7th edition 1993, p651-3

[31] B. Lencová, "Unconventional lens computation", Journal of Microscopy, Vol. 179, Pt. 2, August 1995, p185-190

Appendix 1
Element Integration formulas

1. GAUSSIAN QUADRATURE

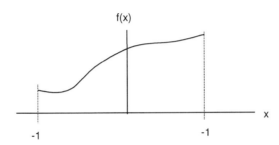

$$\int_{-1}^{1} f(x)dx = \sum_{j=1}^{n} W_j f(a_j)$$ (A1.1)

$\pm a_j$	W_j
n=1	
0	2.00000 00000 00000
n=2	
0.57735 02691 89629	1.00000 00000 00000
n=3	
0.77459 66692 41483	0.55555 55555 55556
0.00000 00000 00000	0.88888 88888 88889
n=4	
0.86113 63115 94053	0.34785 48451 37454
0.33998 10435 84856	0.65214 51548 62546
n=5	
0.90617 98459 38664	0.23692 68850 56189
0.53846 93101 05683	0.47862 86704 99366
0.00000 00000 00000	0.56888 88888 88889
n=6	
0.93246 95142 03152	0.17132 44923 79170
0.66120 93864 66265	0.36076 15730 48139
0.23861 91860 83197	0.46791 39345 72691
n=10	
0.97390 65285 17172	0.06667 13443 08688
0.86506 33666 88985	0.14945 13491 50581
0.67940 95682 99024	0.21908 63625 15982
0.43339 53941 29247	0.26926 67193 09996
0.14887 43389 81631	0.29552 42247 14753

2.   TRIANGLE ELEMENTS

In terms of L_1, L_2 and L_3, the first-order triangle shape functions (triangle co-ordinates)

$$\iint\limits_{Area.} L_1^a L_2^b L_3^c dxdy = \frac{a!b!c!}{(a+b+c+2)!} 2A \qquad (A1.2)$$

where A is the area of the triangle.

Numerical Integration of the function $f(L_1, L_2, L_3)$,

$$F = \sum_{i=1}^{n} W_i f(L_{1i}, L_{2i}, L_{3i}) \qquad (A1.3)$$

	Triangular Co-ordinates			Weights (W_i)
	L_1	L_2	L_3	
Linear	1/3	1/3	1/3	1.0
Quadratic	1/2	1/2	0	1/3
	0	1/2	1/2	1/3
	1/2	0	1/2	1/3
Cubic	1/3	1/3	1/3	-27/48
	0.6	0.2	0.2	25/48
	0.2	0.6	0.2	25/48
	0.2	0.2	0.6	25/48
Quintic	1/3	1/3	1/3	0.225
	α_1	β_1	β_1	0.1323941527
	β_1	α_1	β_1	0.1323941527
	β_1	β_1	α_1	0.1323941527
	α_2	β_2	β_2	0.1259391805
	β_2	α_2	β_2	0.1259391805
	β_2	β_2	α_2	0.1259391805

where

$\alpha_1 = 0.0597158717$
$\beta_1 = 0.4701420641$
$\alpha_2 = 0.7974269853$
$\beta_2 = 0.1012865073$

Appendix 2
Second-order 9 node rectangle element pictorial stars

Rectilinear Coordinates

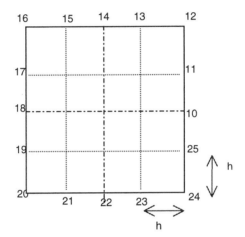

Figure A2.1

The pictorial star at node 1 in Figure A2.1 is given by,

$$
\begin{pmatrix}
-\dfrac{1}{112} & \dfrac{5}{112} & -\dfrac{3}{112} & \dfrac{5}{112} & -\dfrac{1}{112} \\[2mm]
\dfrac{5}{112} & -\dfrac{1}{7} & -\dfrac{9}{56} & -\dfrac{1}{7} & \dfrac{5}{112} \\[2mm]
-\dfrac{3}{112} & -\dfrac{9}{56} & 1 & -\dfrac{9}{56} & -\dfrac{3}{112} \\[2mm]
\dfrac{5}{112} & -\dfrac{1}{7} & -\dfrac{9}{56} & -\dfrac{1}{7} & \dfrac{5}{112} \\[2mm]
-\dfrac{1}{112} & \dfrac{5}{112} & -\dfrac{3}{112} & \dfrac{5}{112} & -\dfrac{1}{112}
\end{pmatrix} u_{ij}
$$

Axisymmetric coordinates

Let R be the radius at the centre node 1 in Figure A2.1. The pictorial star is given by,

$$
\begin{pmatrix}
-\dfrac{h+R}{112R} & \dfrac{5h+5R}{112R} & -\dfrac{3h+3R}{112R} & \dfrac{5h+5R}{112R} & -\dfrac{h+R}{112R} \\[2ex]
\dfrac{2h+5R}{112R} & -\dfrac{h+4R}{28R} & -\dfrac{8h+9R}{56R} & \dfrac{h+4R}{28R} & \dfrac{2h+5R}{112R} \\[2ex]
-\dfrac{3}{112} & -\dfrac{9}{56} & 1 & -\dfrac{9}{56} & -\dfrac{3}{112} \\[2ex]
-\dfrac{2h+5R}{112R} & \dfrac{h-4R}{28R} & \dfrac{8h-9R}{56R} & \dfrac{h-4R}{28R} & -\dfrac{2h+5R}{112R} \\[2ex]
\dfrac{h-R}{112R} & -\dfrac{5h+5R}{112R} & \dfrac{3h-3R}{112R} & -\dfrac{5h+5R}{112R} & \dfrac{h-R}{112R}
\end{pmatrix} u_{ij}
$$

Rectilinear Coordinates

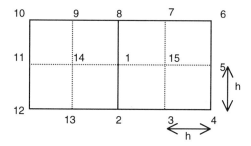

Figure A2.2

The pictorial star at node 1 in Figure A2.2 is given by,

$$
\begin{pmatrix}
\dfrac{5}{176} & -\dfrac{1}{11} & -\dfrac{9}{88} & -\dfrac{1}{11} & \dfrac{5}{176} \\[2ex]
0 & -\dfrac{3}{11} & 1 & -\dfrac{3}{11} & 0 \\[2ex]
\dfrac{5}{176} & -\dfrac{1}{11} & -\dfrac{9}{88} & -\dfrac{1}{11} & \dfrac{5}{176}
\end{pmatrix} u_{ij}
$$

Axisymmetric coordinates

For R at node 2 in Figure A2.2, the pictorial star is given by,

$$\left(\begin{array}{ccccc} \dfrac{8h+5R}{176(h+R)} & -\dfrac{7h+4R}{44(h+R)} & -\dfrac{10h+9R}{88(h+R)} & -\dfrac{7h+4R}{44(h+R)} & \dfrac{8h+5R}{176(h+R)} \\[3mm] 0 & -\dfrac{3}{11} & 1 & -\dfrac{3}{11} & 0 \\[3mm] \dfrac{2h+5R}{176(h+R)} & -\dfrac{h+4R}{44(h+R)} & -\dfrac{8h+9R}{88(h+R)} & -\dfrac{h+4R}{44(h+R)} & \dfrac{2h+5R}{176(h+R)} \end{array} \right) u_{ij}$$

Rectilinear Coordinates

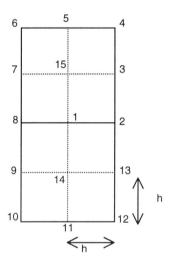

Figure A2.3

The pictorial star at node 1 in Figure A2.3 is given by

$$\left(\begin{array}{ccc} \dfrac{5}{176} & 0 & \dfrac{5}{176} \\[3mm] -\dfrac{h+4R}{44R} & -\dfrac{2h+3R}{11R} & -\dfrac{h+4R}{44R} \\[3mm] -\dfrac{9}{88} & 1 & -\dfrac{9}{88} \\[3mm] \dfrac{h-4R}{44R} & \dfrac{2h-3R}{11R} & \dfrac{h-4R}{44R} \\[3mm] \dfrac{5}{176} & 0 & \dfrac{5}{176} \end{array} \right) u_{ij}$$

Axisymmetric coordinates

For R at node 1 in Figure A2.3 the pictorial star is given by

$$
\begin{pmatrix}
\dfrac{5}{176} & 0 & \dfrac{5}{176} \\[2mm]
-\dfrac{h+4R}{44R} & -\dfrac{2h+3R}{11R} & -\dfrac{h+4R}{44R} \\[2mm]
-\dfrac{9}{88} & 1 & -\dfrac{9}{88} \\[2mm]
\dfrac{h-4R}{44R} & \dfrac{2h-3R}{11R} & \dfrac{h-4R}{44R} \\[2mm]
\dfrac{5}{176} & 0 & \dfrac{5}{176}
\end{pmatrix} u_{ij}
$$

Rectilinear Coordinates

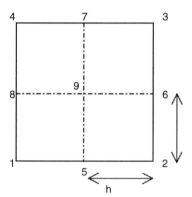

Figure A2.4

The pictorial star at node 9 in Figure A2.4 is given by,

$$
\begin{pmatrix}
-\dfrac{1}{16} & -\dfrac{3}{16} & -\dfrac{1}{16} \\[2mm]
-\dfrac{3}{16} & 1 & -\dfrac{3}{16} \\[2mm]
-\dfrac{1}{16} & -\dfrac{3}{16} & -\dfrac{1}{16}
\end{pmatrix} u_{ij}
$$

Axisymmetric coordinates

For R at node 1 in Figure A2.4 the pictorial star is given by

$$
\begin{pmatrix}
-\dfrac{7h+4R}{64(h+R)} & -\dfrac{4h+3R}{16(h+R)} & -\dfrac{7h+4R}{64(h+R)} \\[2ex]
-\dfrac{3}{16} & 1 & -\dfrac{3}{16} \\[2ex]
-\dfrac{h+4R}{64(h+R)} & -\dfrac{2h+3R}{16(h+R)} & -\dfrac{h+4R}{64(h+R)}
\end{pmatrix} u_{ij}
$$

Appendix 3
Green's Integration formulas

First scalar theorem

$$\iiint_V [a\nabla \cdot (\mu\nabla b) + \mu(\nabla a) \cdot (\nabla b)]dV = \oiint_S a\mu \frac{\partial b}{\partial n}dS \qquad (A3.1)$$

Second scalar theorem

$$\iiint_V [a\nabla \cdot (\mu\nabla b) - b\nabla \cdot (\mu\nabla a)]dV = \oiint_S \mu\left(a\frac{\partial b}{\partial n} - b\frac{\partial a}{\partial n}\right)dS \qquad (A3.2)$$

First Vector theorem

$$\iiint_V [\mu(\nabla \times \mathbf{a}) \cdot (\nabla \times \mathbf{b}) - \mathbf{a} \cdot (\nabla \times \mu\nabla \times \mathbf{b})]dV = \oiint_S \mu(\mathbf{a} \times \nabla \times \mathbf{b}) \cdot \hat{\mathbf{n}}dS$$

$$(A3.3)$$

Second Vector theorem

$$\iiint_V [\mathbf{b} \cdot (\nabla \times \mu\nabla \times \mathbf{a}) - \mathbf{a} \cdot (\nabla \times \mu\nabla \times \mathbf{b})]dV = \oiint_S \mu(\mathbf{a} \times \nabla \times \mathbf{b} - \mathbf{b} \times \nabla \times \mathbf{a}) \cdot \hat{\mathbf{n}}dS$$

$$(A3.4)$$

Appendix 4
Near-axis analytical solution for the solenoid test example

In cylindrical axisymmetric coordinates (r,z), the vector potential $A(r,z)$ and field components $B_r(r,z)$ and $B_z(r,z)$ near the optic axis are given by

$$A(r,z) = \frac{4a^2 \mu_0 Ir}{s^3} f(k)$$

$$B_r(r,z) = \frac{4a^2 \mu_0 Irz}{s^5} \left[3f(k) + k \frac{\partial f}{\partial k} \right]$$

$$B_z(r,z) = \frac{4a^2 \mu_0 I}{s^3} \left[f(k) \left(2 - \frac{3r(a+r)}{s^2} \right) + \frac{\partial f}{\partial k} \left(\frac{k}{2} - \frac{(r+a)k^3}{4a} \right) \right]$$

where

$$f(k) = c_2 + c_3 k^2 + c_4 k^4 + c_5 k^6 + \dots\dots\dots \tag{A4.1}$$

The constants c_j can be evaluated from

$$c_j = a_j - b_j - \frac{a_{j-1}}{2} \tag{A4.2}$$

where a_j and b_j are coefficients involved in the near-axis expansions of the Elliptic functions $K(k)$ and $E(k)$. These expansions are given by

$$K(k) = \frac{\pi}{2} \left[1 + a_1 k^2 + a_2 k^4 + a_3 k^4 + \dots\dots \right]$$

$$E(k) = \frac{\pi}{2} \left[1 + b_1 k^2 + b_2 k^4 + b_3 k^4 + \dots\dots \right]$$

where

$$a_1 = \left(\frac{1}{2}\right)^2$$

$$a_2 = a_1\left(\frac{3}{4}\right)^2$$

$$a_3 = a_2\left(\frac{5}{6}\right)^2$$

$$a_4 = a_3\left(\frac{7}{8}\right)^2 \ldots\ldots a_n = a_{n-1}\left(\frac{2n-1}{2n}\right)^2 \qquad\qquad \text{(A4.3)}$$

$$b_1 = -a_1$$

$$b_2 = -\frac{a_2}{3}$$

$$b_3 = -\frac{a_3}{5} \ldots\ldots b_n = -\frac{a_n}{2n-1}$$

Appendix 5
Deflection fields for a conical saddle yoke in free space

The following derivation has been carried out by Zhao [1]. The geometry of a conical saddle (see Figure A5.1) is defined by its smallest radius R_1, length in z-direction, winding semi-angle ϕ, conical semi-angle α, and number of turns N on the coil. Each coil carries a current I, flowing in the direction indicated in the figure. The derivation here follows Munro's work [2], except that the saddle coils are generalized to a conical shape instead of a cylindrical one and the thickness of the coils is also considered in the final expressions.

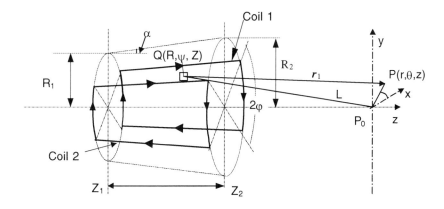

Figure A5.1 A conical saddle deflection yoke in free space.
(Only x-coils which generate the B_x field are shown)

Let P be a general point near the deflection axis, with cylindrical coordinates (r, θ, z); let Q be general point, with cylindrical coordinates (R, ψ, Z), lying on the conical surface S bounded by each coil, and let dS be an element of area on the conical surface in the neighborhood of point Q. The derivation will be based on the Biot-Savart law expressed in terms of scalar potential Φ, defined as H=-∇Φ, so

$$\Phi = -\frac{NI}{4\pi}\iint_S \frac{\mathbf{r}_1 \times \mathbf{n}dS}{|\mathbf{r}_1|^3} \tag{A5.1}$$

where NI is the ampere-turns of each coil, $\mathbf{n}$ is the unit vector normal to the surface element dS

$$\mathbf{n} = \cos\alpha\cos\psi\,\mathbf{i} + \cos\alpha\sin\psi\,\mathbf{j} + \sin\alpha\,\mathbf{k} \tag{A5.2}$$

and $\mathbf{r}_1$ is the vector from dS to P

$$\mathbf{r}_1 = (r\cos\theta - R\cos\psi)\mathbf{i} + (r\sin\theta - R\sin\psi)\mathbf{j} + (z - Z)\mathbf{k} \tag{A5.3}$$

here R can be expressed as a linear function of Z

$$R = (Z - Z_1)\tan\alpha + R_1 \tag{A5.4}$$

The potential Φ_1 at point P produced by coil 1 is calculated from equation (A5.1)

$$\Phi_1 = \frac{NI}{4\pi}\iint_S \frac{[R - r\cos(\psi - \theta)]\cos\alpha - (z - Z)\sin\alpha}{\left[r^2 - 2rR\cos(\psi - \theta) + R^2 + (z - Z)^2\right]^{\frac{3}{2}}}\frac{R}{\cos\alpha}dZd\psi \tag{A5.5}$$

by defining $L = \sqrt{R^2 - (z - Z)^2}$,

$$\Phi_1 = \frac{NI}{4\pi}\iint_S \frac{R[R - r\cos(\psi - \theta)]\cos\alpha - R(z - Z)\sin\alpha}{L^3\cos\alpha}$$

$$\left[1 - \frac{2rR}{L^2}\cos(\psi - \theta) + \frac{r^2}{L^2}\right]^{-\frac{3}{2}}dZd\psi \tag{A5.6}$$

Since $|r| \le L$, the square-bracketed term can be expanded by the binomial theorem, to give

$$\Phi_1 = \frac{NI}{4\pi\cos\alpha}\iint_S \frac{R[R - r\cos(\psi - \theta)]\cos\alpha - R(z - Z)\sin\alpha}{L^3}$$

$$\left[1 + \frac{3rR}{L^2}\cos(\psi - \theta) - \frac{3r^2}{2L^2} + \frac{15r^2R^2}{2L^4}\cos^2(\psi - \theta)\right.$$

$$\left. -\frac{15r^3R}{2L^4}\cos(\psi - \theta) + \frac{35r^3R^3}{2L^6}\cos^3(\psi - \theta) + ...\right]dZd\psi$$

$$\tag{A5.7}$$

Similarly, the potential Φ_2 produced by coil 2 is found to be

$$\Phi_2 = \frac{NI}{4\pi\cos\alpha} \iint_S \frac{R[R + r\cos(\psi-\theta)]\cos\alpha - R(z-Z)\sin\alpha}{L^3}$$

$$\left[1 - \frac{3rR}{L^2}\cos(\psi-\theta) - \frac{3r^2}{2L^2} + \frac{15r^2R^2}{2L^4}\cos^2(\psi-\theta)\right.$$

$$\left. + \frac{15r^3R}{2L^4}\cos(\psi-\theta) - \frac{35r^3R^3}{2L^6}\cos^3(\psi-\theta) + ...\right] dZ d\psi$$

$$(A5.8)$$

The total potential Φ produced by the entire yoke can be obtained by adding eq.(A5.7) and (A.5.8) together

$$\Phi = \Phi_1 + \Phi_2$$

$$= \frac{NI}{2\pi} \iint_S \left\{ \frac{Rr}{L^3}\left[-1 + \frac{3R[R-(z-Z)\tan\alpha]}{L^2}\right]\right.$$

$$+ \frac{Rr^3}{L^5}\left[\frac{3}{2} - \frac{15R[R-(z-Z)\tan\alpha]}{2L^2}\right.$$

$$\left. - \frac{45R^2}{8L^2} + \frac{105R^3[R-(z-Z)\tan\alpha]}{8L^4}\right]\right\}\cos(\psi-\theta)$$

$$+ \frac{r^3R^3}{L^7}\left[-\frac{15}{8} + \frac{35R[R-(z-Z)\tan\alpha]}{8L^6}\right]\cos 3(\psi-\theta) + ... \right] dZ d\psi$$

$$(A5.9)$$

Assuming $G = R[R-(z-Z)\tan\alpha]$

$$\Phi = \frac{NI}{2\pi}\iint_S \left\{\left[\frac{Rr}{L^3}\left(-1+\frac{3G}{L^2}\right)\right.\right.$$

$$\left.\left. + \frac{Rr^3}{L^5}\left(\frac{3}{2} - \frac{45R^2+60G}{8L^2} + \frac{105R^2G}{8L^4}\right)\right]\sin\psi\cos\theta\right.$$

$$\left. + \frac{r^3R^3}{L^7}\left(-\frac{5}{8} + \frac{35G}{24L^6}\right)\sin 3\psi\cos 3\theta + ...\right\} dZ$$

$$(A5.10)$$

By comparing equation (A.5.10) with the first and third harmonic deflection field functions, $d_1(z)$ and $d_2(z)$, are found to be

$$d_1(z) = \frac{NI}{\pi} \sin\varphi \int_{Z_1}^{Z_2} \frac{R}{L^3}\left(\frac{3G}{L^2} - 1\right) dZ$$

$$d_3(z) = \frac{15NI}{8\pi} \sin 3\varphi \int_{Z_1}^{Z_2} \frac{R^3}{L^7}\left(\frac{35G}{L^2} - \frac{5}{8}\right) dZ$$

(A5.11)

In the above derivation, the coils are assumed to be lying on a conical surface with no thickness. To extend these formulas to a coil of finite thickness

$$d_1(z) = \frac{NI}{\pi\Delta R} \sin\varphi \int_{Z_1}^{Z_2} \int_{R}^{R+\Delta R} \frac{R}{L^3}(\frac{3G}{L^2} - 1) dRdZ$$

$$d_3(z) = \frac{15NI}{8\pi\Delta R} \sin 3\varphi \int_{Z_1}^{Z_2} \int_{R}^{R+\Delta R} \frac{R^3}{L^7}(\frac{35G}{L^2} - \frac{5}{8}) dRdZ$$

(A5.12)

where ΔR is the thickness of the coils, and R, described by equation (A5.4) , denotes the innermost linear equation defining the winding profile of the saddle coils between Z_1 and Z_2.

REFERENCES

[1]   Y. Zhao, "The design of In-Lens deflectors for electron beam systems", PHD thesis, Electrical Engineering Department, National University of Singapore, Singapore 119260, to be published

[2]   Munro, E. and Chu, H.C. Numerical analysis of electron beam lithography system. Part I: computation of fields in magnetic deflectors, Optik, 60, pp.371-390. 1982

INDEX

A

Aberrations 32
Accuracy tests
    Closed Tube 58
    Electric box 48, 239
    Parabolic capacitor 247
    Solenoid coil 164
    Spherical capacitor 222, 253
    S100 objective lens 177
    Saturated objective lens 181
    Two-tube lens 141
Ampere's circuital law 11
Ampere's circuital mesh test 248
Axisymmetric problems 27, 56, 92

B

Bio-Savart's law 19
Brick element 112
Boundary conditions
    Derivative 4
    Explicit 4
    Material Interfaces 8, 12
Boundary element method 202
Boundary-fitted coordinate Method 217

C

C^1 triangle interpolant 231
Chromatic aberration 33
Conjugate gradient (CG) 49
Conservative property 1
Cubic elements 105
Current loading function 40, 42

D

Delaunay triangle mesh 209
Diffraction limit 33
Direct ray tracing 154, 185
Dirichlet boundary conditions 4
Divergence theorem 7

E

Edge elements 96
Electron Guns 26, 224, 246
Electron lenses
    Magnetic 28, 161
    Electric 45, 141

F

Field expansions 30
Field interpolation 157, 230
Five pointed star 46

G

Galerkin's method 84
Gauge condition 22
Gaussian elimination 49
Gaussian Quadrature integration 257
Gaussian plane 32
Gauss's law 2
Graded mesh 143, 183, 249
Green's theorems 265

H

Harmonic scalar potential 39
High-order elements
    Triangle 100
    9 node quadrilateral 105
    Serendipity family 106
High-order interpolation
    C^1 triangle interpolant 231
    Normalised Hermite-Cubic 232
    Laplace polynomial Interpolant 234

I

ICCG 49, 60
Interpolation error 230

J

Jacobian, 84, 218, 232

K

Khursheed Electron Optics Software (KEOS) 155, 195, 212

L

Laplace's equation 4
Laplace polynomial expansion 234
Lorentz force 25

M

Magnetic scalar potential 17
Matrix equation 47
Mesh refinement 209
Mesh generation 209
Multipole electrodes 43

N

Nine pointed star 55, 59
Numerical integration formulas 257
Nodal equation 89
Node numbering 87
Neumann boundary conditions 5

O

Octopole deflector 38

P

Parabolic capacitor 247
Paraxial equation 29
Poisson's equation 4
Positive definite 66

Q

Quasi-conformal mesh 183, 212
Quasi-curvilinear mesh 212, 226
Quadrilateral Elements 82
Quadrupole structure 219

R

Reduced scalar potential 18
Region Block Mesh 143, 176, 206
Residual error 62

S

Saddle Coil 38
Saturation 12, 135
Serendipity Elements 106

Second-order Elements
  9 node 105, 259
  Quadrilateral 105
  Triangle 99
  Serendipity 106
Self-adjoint 66
Shape functions
  One dimensional 76
  Two dimensional 77
  Three dimensional 112
  Vector 96
Square elements 70
Solenoid coil
  Off-axis field solution 164
  Near-axis field solution 267
Secondary electrons 27
Spherical aberration
  Third-order 32
  Fifth-order 155, 185
Spurious solutions 96
Stiffness matrix 90
Stormer's potential 29
Structured mesh 209
Successive Overrelaxation 49

T

Tetrahedral elements 113
TOSCA program 187
Toroidal Coil 38
Total scalar potential 20
Triangle Elements 76
Triangle integration formulas 258
Truncation error 49, 159
Two-tube lens 141

U

Unstructured mesh 209

V

Variational principle 66
Vector Potential 13

W

Weighted residual integral 62